The Hoopa Project

Bigfoot Encounters in California

Dedication

There is no possible way that I could ever thank these two people for their never-ending support, positive ideas, great attitude, and generally standing by me when I was down. This book could never have been completed without you two continually being my cheerleaders and rooting section. I couldn't have asked for two better people to help me in so many different ways. I love you both so much that I can't put it into words. Here's a big hug and thank you to Ben and Nicole! This book is dedicated to you!

Dad

The Hoopa Project

Bigfoot Encounters in California

By David Paulides

ISBN-13 978-0-88839-653-2

2nd printing 2008

Cataloging in Publication Data

Paulides, David
The Hoopa project : bigfoot encounters in California / by David Paulides.

Includes index.
ISBN 978-0-88839-653-2

1. Sasquatch—California—Humboldt County. I. Title.

QL89.2.S2P385 2008 001.944 C2008-902846-5

Printed in China — Jade Productions

Production: Mia Hancock
Cover Design: Ingrid Luters
Front cover illustration: Harvey Pratt

Published simultaneously in Canada and the United States by

HANCOCK HOUSE PUBLISHERS LTD.
19313 Zero Avenue, Surrey, B.C. Canada V3S 9R9
(604) 538-1114 Fax (604) 538-2262

HANCOCK HOUSE PUBLISHERS
1431 Harrison Avenue, Blaine, WA U.S.A. 98230-5005
(604) 538-1114 Fax (604) 538-2262

***Website:* www.hancockhouse.com**
***Email:* sales@hancockhouse.com**

Contents

God Rest Their Souls

Sergeant Joe Masten *Hoopa Tribal Police*
Phil Smith *Bluff Creek Resort*
Tony Hacking *United States Forest Service*
Pliny McCovey *Hoopa Tribal Member*

I have been fascinated by stories of Bigfoot, Sasquatch, Chinese Wildmen, [the] Abominable Snowman from all around the world for a very long time; and I have been corresponding with many people, including a number who have allegedly seen one of these beings, and I've always had, I suppose, a romantic outlook. I would like to believe that these creatures exist. The fact they they are talked about in so many parts of the world really makes me think that there is something there — there must be...from Ecuador to China to Russia and to different parts of the ex-Soviet Union and all over North America...Native Americans I know have actually seen Sasquatch, they believe...and very respected people who've at least heard something which to them sounds like Sasquatch or Bigfoot.

Jane Goodall
Primatologist
Taped November 1, 2003
Bigfoot Research Symposium 2003
Pacific Science Center, Seattle, Washington

Introduction

Knowledge of our own arrogance is the first step toward true knowledge.

SOCRATES

In 1968 I was twelve years old and indoctrinated into my dad's lifestyle. We spent as much free time as possible walking the woods of Northern California, fishing, sightseeing and enjoying wildlife. We never went hunting, but my dad's past profession as a police officer prompted him to always carry a gun while on our trips. We spent many summers backpacking into the far reaches of Trinity, Humboldt, Shasta, Lassen, and Modoc County. On past deer-hunting trips, my dad and grandfather had explored these areas and knew them well prior to letting me tag along. We had probably hiked into the South Antelope Creek region of Lassen County five times over an eight-year span, but the sixth time was different, very different.

It was in August and the new school year was quickly approaching. The ride into the region was dusty, hot and extremely dry. I remember that we always passed a California Department of Corrections prison; somehow this always stuck out in my mind. We drove through a creek and made our way to a dirt overlook. We parked our car, put our packs on, and started the two-hour dusty and hot hike down into the creek. There was always anticipation in getting to our fishing spot, as the fishing was always exceptional. In over twenty trips down into the creek, my dad had never seen another person in the area and had never seen any indicators that people had been there. You never wanted to hike out of the valley in the middle of the day due to the heat, snakes, and the lack of water. The hike down into the creek was fairly easy, while the hike out was brutal.

We arrived at our camping spot next to the creek where we always stopped and slept on previous trips. There was a large umbrella of huge trees that kept the creek and the surrounding area cool and refreshing. There was a shallow cutout where the river had gouged a small cave into the hillside where we slept. We started to set out our equipment and put our poles and reels together when we started to smell smoke. Smoke in this area is a very dangerous issue because fire can roar through the valley quickly. We knew that there was nobody in the valley because we were at the only parking location and there wasn't another car. If this wasn't a controlled burn, then fire could quickly develop and trap us in the valley.

My dad told me to put my pole down and walk with him upstream towards the smoke. We made our way quickly up the creek and started to close in on the smoke. I made a mental note that the creek was just small enough that I could probably cross in certain spots. I also noted that there was no garbage, footprints or other tracks in the sand bank, or other signs of humans. It was a very serene location; but as we got closer to the smoke I could tell by my dad's mannerisms that he was uncomfortable with what was happening.

We worked our way through some large spruces, which immediately opened up to a beach area. The sand area was approximately thirty feet square in size. The creek had flowed off several large boulders into a deep pool that was adjacent to the beach. You could see where the creek had washed a few feet up onto the sand and see where the sand had changed colors because of the moisture content. It was just inside this discoloration zone we saw the fire. It is something I will never forget.

The fire was made of fifteen to twenty small twigs broken into equal pieces approximately one foot long. They were placed on the sand in the shape of a teepee. As we walked up to the fire my dad got very nervous. He pulled his gun and scanned the surrounding area. He slowly made his way towards the fire while also looking at the sand in the surrounding area. Well, being the son of my father, I was also pretty nervous and was looking at the sand. I immediately noticed how flat the sand was, no ripples, footprints, nothing. The area where the river had changed the color of the sand was dead level, no marks anywhere. The surrounding area was absolutely quiet except for the creek — eerily quiet. There were no birds chirping, no squirrels running around, no sound except the creek. It was almost as if a vacuum had sucked all of the noise out of the atmosphere, except for the water running over the boulders.

My dad had spent three years with San Francisco Police Department and then transferred to the fire department. He was in the middle of a thirty-year career in public service when we made this trip. I think he was half police officer and half arson inspector when we were on that beach. He knelt next to the fire and slowly and methodically pushed the small twigs down looking for the ignition source. He couldn't find anything. We continued to scan the beach for indications of footprints, sled marks, bicycle tracks (ridiculous) — heck anything — but nothing was visible. My dad put the fire out, kept his gun out, and told me to start walking back towards our gear.

After several minutes we arrived back at our packs. My dad told me that he was extremely uncomfortable staying down in the valley overnight because it was obvious that something or someone else was living in this valley and they wanted to scare us off. They had succeeded, and we were leaving. As we collected our belongings, he continued to look behind us towards the direction of the fire and kept his gun out. This was the first time I had ever seen my dad as nervous and protective as he was at that very minute. His actions made me nervous, but I also knew that he was a rational thinker and it was in our best interests if we got out of the valley.

The sun was lying low on the horizon and was casting shadows, making the hike out a lot less brutal than it would have been in the heat of the day. A normal hike out from the South Antelope at that time of the day would take us four hours; we made it in two. There was little said between us other than sharing canteens. We got back to the car very exhausted and drove back into Redding. Only when we reached Interstate 5 for the six-hour drive back to our house did my dad show signs of relaxing.

We stopped at an A & W Root Beer stand in Redding to buy hamburgers. We placed our order and were sitting in the darkness looking back towards Mt. Lassen as the moon was casting light on the side of its peak. My dad said that he had been visiting those woods for a large portion of his life, and he had seen hunters, other fishermen, and even dead animals in his travels. He told me that what we had just witnessed was the most peculiar and uncomfortable thing he had witnessed in his entire life. He felt it was obvious that someone, or something, with intelligence started the fire, and that we could assume that the fire was started just as we arrived in the valley, as a message to make us concerned. He explained that we should have seen tracks near the fire, and that there should have been an ignition source. When he couldn't find tracks or an ignition source, he decided we weren't staying there. He admitted that he had no idea who or what made the fire, and that we might never find out.

During the ride home we talked about a variety of animals that were in the forest, and how the fire could possibly had been started. This was the first time I ever heard of Bigfoot. My dad said that hunters and fishermen throughout Northern California had seen Bigfoot for years. He described Bigfoot, how it looked, and explicitly stated that Bigfoot had never injured anyone. While he had read stories of Bigfoot doing mischievous things to make visitors leave an area it inhabited, he had never heard of Bigfoot using fire. He did confirm that Bigfoot left human-like tracks that were easily visible, and that the local Native American Indians had stories of Bigfoot going back hundreds of years. He had never told me about this animal, he said, because he never wanted me to be scared, and then apologized for even bringing it up, because he knew that Bigfoot couldn't have made the fire.

We arrived home long after midnight. I think I slept till noon the next day and staggered into the kitchen to hear my parents talking about our journey. Mom was staring in disbelief at Dad's story. My dad had called the Department of Correction (DOC) facility near the valley and asked if anyone had escaped in the recent past. The DOC stated that they had never had anyone escape that wasn't immediately recaptured, and nobody had left in several months.

I still think through what happened in that valley that day. While I truly don't believe that Bigfoot lit that fire, I do believe that there was some intelligent being in the area that did. The fire was too neatly constructed for it not to be made by something smart. In my years talking to the Hoopa, Yurok, and Kurok Native Americans I did learn a lot about the types of creatures, people, and superstitions that exist in the mountains of Northern California. Maybe our fire starter was a "little person," a being that has been described by many Native Americans as small humans that live in caves and underground, and come out at night to roam the forests of Northern California. (The Karuks believe they are descended from little people.) Maybe it was a human who was living off the grid and didn't want us infringing on the good thing he had at the creek. Maybe the incident is even more sinister than I could imagine. Who knows?

Second Incident

The second incident involving my personal education about Bigfoot occurred much later in life. I was now the father and was with my daughter. In July 2002 I was in Whistler, British Columbia with my ex-wife, son, and daughter for my son's hockey tournament. His team had just won the California State Pee Wee hockey championship and they were invited to Whistler to play in a North American tournament. We were there for five days of hockey, which equaled five days of hell for my daughter. She was ten years old at the time and a very patient sister, but a sister looking for some excitement. One of my son's games was completed at 11 a.m. and I agreed to take my daughter on a fishing trip into the mountains surrounding Whistler; she loved fishing. I contacted a local guide and we met him at his shop. The owner made the introductions to our guide, Curtis Christian. Little did I know at the time, the introduction to Curtis would change my life.

Curtis was in his late twenties, tanned, in great shape, and a very humble and polite personality. He was a perfect match for my daughter and me. Curtis asked what type of fishing we liked, how advanced my daughter was with a rod, and what type of fish we liked to catch. We had a short conversation, loaded the equipment in Curtis' truck and made our way into the mountains.

The hills surrounding Whistler are large, rugged, and absolutely beautiful. There are many streams that run down the valleys into the Whistler region, making several options for fishing. The three of us sat in the front seat of Curtis' truck and talked stories for our ninety-minute ride. Curtis told me that he had been coming to Whistler to spend his summers for many years. He said that he loved the wildlife, scenery, and fishing. I am always inquisitive about a person's life and asked Curtis what he did in his winters. He was a coach for the Canadian Olympic Team.

We were now on a very rough dirt road that took Curtis' four-wheel drive everything it had to make it up and over some very large boulders. We were following a small creek up a ravine when I asked Curtis if he had ever had any strange, outdoor wildlife experiences while guiding. He looked directly in my eyes and then down at my daughter and gave me the type of look questioning whether he could talk in front of my daughter. I said that he could talk freely, as my daughter had been in some of the most remote areas of North America and was very comfortable in the outdoors.

Curtis said that the previous summer he was guiding two clients for a day of fishing similar to what he was doing with us. The situation was also quite similar, driving up a road adjacent to a small river with a lot of brush cover near the bank, probably fifteen miles from anyone or any civilization. They were driving fairly slowly when they all saw something walk out in front of them from the side of the road next to the river. It was early in the morning but the sun was fully up and the visibility was excellent. He told me that they saw a huge, hair-covered animal walk directly in front of them into the middle of the road. It walked on two feet, never on four, and had the posture and walking strut of a normal person, and the face almost of a human being.

Curtis was looking in my eyes for my expression or my reaction, but I had none. I was mesmerized. He said that the animal looked back at them and took two huge strides and disappeared into the brush and timber on the opposite side of the road. He said he knew he had just seen Sasquatch. My daughter had heard stories of Bigfoot and was also very interested in the story, and wasn't scared. Curtis continued to offer additional details of the sighting until we reached the alpine lake we would be fishing. Curtis is the kind of person who was polite enough to not want to scare my daughter, but certain enough about what he saw that he wasn't looking for my nod of approval or acceptance.

My daughter and I used float tubes the entire day to fish in the lake, caught few fish, but had a great father–daughter day. I must admit I wasn't thinking much about the fish that day, but more about Bigfoot.

This Whistler incident prompted me to read everything I could find about Bigfoot. This was just after the Patterson–Gimlin film was re-digitized, and there were still many articles being written about Bigfoot. Between school, sports, work and lifestyle issues, however, Bigfoot slowly moved to the back of the file cabinet and I moved forward with my life. It wasn't until my lifestyle became comfortable, money was in the bank, and kids were in school that I found the time to wade into the Bigfoot issue. The following is the story of my California Bigfoot Search, spotlight: Hoopa.

David Paulides

Foreword

This remarkable work brings us another first in the annals of Bigfoot research.

Numerous Bigfoot-related incidents are presented that are the result of first-hand investigations by a former trained police officer and professional investigator. Added to this, a highly qualified, experienced forensic police artist was commissioned to meet Bigfoot sighting witnesses and to prepare detailed sketches of what the witnesses saw.

David Paulides recorded his conversations with Bigfoot witnesses in the same way he prepared his reports in police work. The entire book is therefore written in what might be termed "police report format," which is very direct and to the point. Particular care is taken to ensure that facts are not misread or misinterpreted. Sentences are short and thoughtfully structured to ensure the reader fully comprehends what is being said.

The details, interpretations, and images provided on Bigfoot encounters in this book go far beyond "normal reporting." Dave Paulides brings to the table over 30 years of professional investigative experience, and we are therefore provided with what can only be termed as the best Bigfoot reports ever written.

Christopher L. Murphy
(Author, *Meet the Sasquatch*)

Chapter 1
Government Acknowledgement

When I was completing my graduate degree I became comfortable compiling masses of information and delving into issues to find hidden agendas, etc. With my thesis finished I knew that I was well prepared to tackle any big project that came my way. The techniques I'd learned were quite helpful when I moved into the technology sector as a vice president of business development, responsible for worldwide business alliances associated with the laser industry and database development. I was able to use my skills and abilities to push forward corporate agendas. I always had an easy time ciphering out corporate issues and getting information on companies; it is the government that has historically been secretive about releasing sensitive information to the public.

The Mission

As someone who worked for municipalities for twenty years as a police officer, I have a deep appreciation for what governments bring to the table. The resources that governments at all levels can bring to any investigation are substantial. If one level of bureaucracy doesn't have the specific resource, they make a mutual-aid call and help arrives.

Prior to actually going into the field looking for a suitable location to search for Bigfoot and interviewing witnesses, I thought it would be prudent to see if the government has ever acknowledged the existence of Bigfoot. This wasn't an easy task. Most governments don't make public spectacles of themselves, and most elected officials do not want to be embarrassed. I was, however, able to find a few very interesting government documents that were applicable to Bigfoot.

As almost all citizens in any country know, governments avoid putting themselves in awkward situations where they can't defend their positions. Government employees do not want to cause turmoil or embarrassment for themselves or their superiors for fear of reprisals, and, therefore, are among the most cautious of workers. I think we could extrapolate that military leaders and their subordinates are extremely cautious on the public stance they take on issues, and usually have all printed material for public review or inspection approved by several layers of bureaucrats. The Army Corps of Engineers is among the most cautious of all military branches, being used for many key governmental functions throughout the United States and the world. Most recently the Corps was used in New Orleans to stem the flow of water from hurricane Katrina. The Corps is an extremely competent, intelligent and technical part of the U.S. Army.

"The United States Army Corps of Engineers (USACE) is made up of approximately 34,600 civilian and 650 military men and women. Our military and civilian engineers, scientists and other specialists work hand in hand as leaders in engineering and environmental matters. Our diverse workforce of biologists, engineers, geologists, hydrologists, natural resource managers and other professionals meets the demands of changing times and requirements as a vital part of America's Army." (Quoted from the Army Corps of Engineers website.)

In the realm of Bigfoot research we are always looking for credible validation of beliefs. The idea that Bigfoot exists is always the number one issue that we are searching for and attempting to document. In my personal search of the United States government records to acknowledge publicly the issue of Bigfoot, I was successful. The United States Army Corps of Engineers authored an atlas in 1975 on the State of Washington. The atlas encompassed a variety of topics in which the Corps has expertise. Many of the problems the Corps covers represent issues and barriers they may encounter enroute to their mission. One portion of the atlas is a two-page report on Sasquatch. The narrative follows.

> SASQUATCH The very existence of Sasquatch, "BigFoot" as it is sometimes known, is hotly disputed. Some profess to be open-minded about the matter, although stating that not one piece of evidence will withstand serious scientific scrutiny. Others, because of a particular incident or totality of reports over the years, are convinced that Sasquatch is a reality. Alleged Sasquatch hair samples inspected by F.B.I. laboratories

> resulted in the conclusion that no such hair exists on any human or presently known animal for which such data are available.
>
> Information from alleged sightings, tracks and other experiences conjures up the picture of an ape-like creature standing between 8 and 12 feet tall, weighing in excess of 1,000 pounds, and taking strides of up to 6 feet. Plaster casts have been made of tracks showing a large, squarish foot 14 to 24 inches in length and 5 to 10 inches in breadth. Reported to feed on vegetation and some meat, the Sasquatch is covered with long hair, except the face and hands, and has a distinctly humanlike form. Sasquatch is very agile and powerful, with the endurance to cover a vast range in search of food, shelter and others of its kind. It is apparently able to see at night and is extremely shy, leaving minimal evidence of its presence. Tracks are presently the best evidence of its existence. A short film of an alleged female Sasquatch was shot in Northern California, which, although scoffed at, shows no signs of fabrication.
>
> The Pacific Northwest is generally considered to be the hotbed of Sasquatch activity, with Washington leading in numbers of reports of tracks or sightings since 1968. However, reports of Sasquatch-like creatures are known from as far as the Parmir Mountains in the U.S.S.R. and South America.
>
> If Sasquatch is purely legendary, the legend is likely to be a long time dying. On the other hand, if Sasquatch does exist, then with the Sasquatch hunts being mounted and the increasing human population it seems likely that some hard evidence may be soon at hand. Legendary or actual, Sasquatch excites a great popular interest in Washington. 1975, Army Corps of Engineers Washington Atlas.

The narrative also includes a map that shows sighting locations and a small drawing of a Sasquatch. I've included the above listed narrative for assistance in the reading of their document. The fine people at the Corps of Engineers in Seattle e-mailed me a scanned copy of the document and it's included on page 20. The scan didn't copy perfectly so I transcribed the narrative. The sightings map they have included listed the following:

	Tracks	Sightings	Both	Total
British Columbia	38	80	11	138
Washington	32	51	12	95
Oregon	15	25	6	46
California	82	59	10	151
Other	7	42	18	67
Totals	174	266	57	497

In a review of the Corps narrative it is important to note the first paragraph and the statement regarding hair samples. It states that hair samples allegedly from a Sasquatch came back from analysis as not known to any human or animal. This statement is quite important. If the fiber wasn't hair, it would state it was synthetic, cotton, etc. It was hair, but it cannot be identified using any known DNA profile, per the FBI.

The second paragraph describes the creatures within the physical parameters of what we have identified in Hoopa. It also states that they have found no fabrication of the Patterson–Gimlin footage of Bigfoot/Sasquatch. If you believe in your government and you believe they have the best evidence, laboratories, and tools available, then I believe the government is taking a bold step forward in the authentication of Bigfoot/Sasquatch as a creature.

Skamania County, Washington

Skamania County is located in the State of Washington, approximately sixty miles east of Vancouver, Washington at the mouth of the Columbia River gorge. This area is extremely rural, with Gifford Pinchot National Forest to the north and Mount Hood National Forest to the south. Skamania County has a total population of 10,664 as of 2005. The County has 1,672 square miles in area. The largest city in Skamania and the county seat is Stevenson with a population of 1,200.

Skamania County received national recognition in 1969 when the County Commissioners received several reports of Bigfoot sightings. The

SASQUATCH

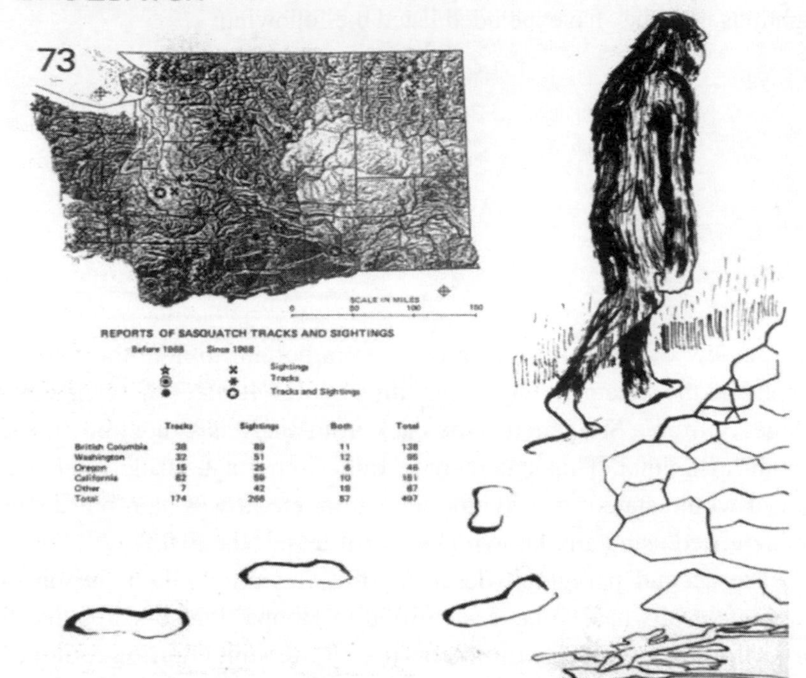

	Tracks	Sightings	Both	Total
British Columbia	38	89	11	138
Washington	32	51	12	95
Oregon	15	25	6	46
California	82	59	10	151
Other	7	42	18	67
Total	174	266	57	497

SASQUATCH

The very existence of Sasquatch, or "Big Foot" as it is sometimes known, is hotly disputed. Some profess to be open-minded about the matter, although stating that not one piece of evidence will withstand serious scientific scrutiny. Others, because of a particular incident or totality of reports over the years, are convinced that Sasquatch is a reality. Alleged Sasquatch hair samples inspected by F.B.I. laboratories resulted in the conclusion that no such hair exists on any human or presently-known animal for which such data are available.

Information from alleged sightings, tracks and other experiences conjures up the picture of an ape-like creature standing between 8 and 12 feet tall, weighing in excess of 1,000 pounds, and taking strides of up to 6 feet. Plaster casts have been made of tracks showing a large, squarish foot 14 to 24 inches in length and 5 to 10 inches in breadth. Reported to feed on vegetation and some meat, the Sasquatch is covered with long hair, except for the face and hands, and has a distinctly humanlike form. Sasquatch is very agile and powerful, with the endurance to cover a vast range in search of food, shelter and others of its kind. It is apparently able to see at night and is extremely shy, leaving minimal evidence of its presence. Tracks are presently the best evidence of its existence. A short film of an alleged female Sasquatch was shot in northern California which, although scoffed at, shows no indications of fabrication.

The Pacific Northwest is generally considered to be the hotbed of Sasquatch activity, with Washington leading in number of reports of tracks or sightings since 1968. However, reports of Sasquatch-like creatures are known from as far away as the Pamir Mountains in the U.S.S.R. and South America.

If Sasquatch is purely legendary, the legend is likely to be a long time in dying. On the other hand, if Sasquatch does exist, then with the Sasquatch hunts being mounted and the increasing human population it seems likely that some hard evidence may soon be in hand. Legendary or actual, Sasquatch excites a great popular interest in Washington.

"MYTHICAL CREATURES SHOULDN'T BE MENTIONED IN THE ATLAS!"

53

commissioners became concerned, and believed that there was a creature matching Bigfoot's description that lived in the rural sections of the county. Their concern prompted the Council to enact legislation, ordinance 69-01.

To understand what may have prompted the county to enact the legislation, I checked the BFRO.net website under their sightings section. I found that Skamania County had the largest number of reported Bigfoot sightings in Washington — fifty. The year range for sightings in Skamania County is 1950–2006.

The first document listed in the following pages is the first piece of legislation that was passed in 1969. In summary, the council states, "there is evidence to indicate the possible existence in Skamania County of a nocturnal primate mammal." This is a monumental declaration for a legislative body to state. The bill goes on to state that any wanton killing of the creature is a felony. See the bill for the exact wording.

In 1984 Skamania County opted to update their Bigfoot legislation adopted in 1969. The update was extensive. The legislation went from one page to three. What follows, on pages 24 to 26, is the entire three-page bill. When I originally called the county councilors, they appeared quite surprised that I had known about the amendment, and specifically asked if I knew the legislation number. They didn't. Here it is for your review.

Skamania County ordinance 1984-2 amends and repeals portions of bill 1969-1. The current legislation talks about the existence of the creature, and the lack of local and national laws to protect it. It then states that the council may have exceeded its authority by making the violation a felony. The council declares its county to be a Sasquatch refuge and describes the creature as "Sasquatch, Yeti, Bigfoot or Giant Hairy Ape." The council further states any type of killing of the creature will be a misdemeanor punishable according to the type of killing, accidental, premeditated, etc.

The most interesting section of the Skamania County legislation is found in section 4. It states, "Should the Skamania County Coroner determine any victim/creature to have been humanoid the prosecuting attorney shall pursue the case under existing laws pertaining to homicide. Should the coroner determine the victim to have been an anthropoid (ape-like creature) the prosecuting attorney shall proceed under the terms of this ordinance."

A county in the United States of America is so concerned about the welfare of a creature that science will not acknowledge exists that it enacts legislation addressing the point of killing a human versus an anthropoid. Fascinating.

Whatcom County, Washington

An Internet search for Bigfoot/Sasquatch legislation came up with Whatcom County, Washington. I called the county council and was fortunate to be assisted by a very friendly county employee. I explained my request and asked if she had ever heard of the legislation. She stated that she had, but had never received a request for a copy. She said that she would gladly send me a certified copy. Their legislation was enacted on June 9, 1991. The bill calls for Whatcom County be declared a "Sasquatch protection and refuge area."

Whatcom County is also located in the State of Washington. Whatcom is the north-westernmost county in the state and sits adjacent to the Canadian border. The county has a population of 183,000 people with a landmass of 2,119 square miles. The county seat is Bellingham, with a population of 61,000 people. Whatcom County has a designation as a National Scenic Byway from Bellingham to Mount Baker (Highway 542). There are locations of Whatcom County in the Mount Baker region that are extremely rural.

In 1992 Whatcom County received national prominence when their county council enacted legislation defining the county as a "Sasquatch Protection and Refuge Area." The legislation was passed and approved on 6/9/91 and was formally implemented in 1992, thus the date of the bill.

A check of Bigfoot Internet databases lists forty sightings in Whatcom County, which makes its rank as eighth on the Washington County list for the most Sasquatch/Bigfoot sightings. Five of the sightings listed on the site are near the Mount Baker/Lake Baker area. The years that sightings have been reported range from 1972–2005. Whatcom County did not list any penalties for violating the legislation, and the language in the bill is very limited as compared to Skamania County's bill. The legislation is written on a single page and it is shown on page 27.

Whatcom County has not amended or repealed the legislation and it still stands as county law.

ORDINANCE NO. 69-01

BE IT HEREBY ORDAINED BY THE BOARD OF COUNTY COMMISSIONERS OF SKAMANIA COUNTY:

WHEREAS, there is evidence to indicate the possible existence in Skamania County of a nocturnal primate mammal variously described as an ape-like creature or a sub-species of Homo Sapian; and

WHEREAS, both legend and purported recent sightings and spoor support this possibility, and

WHEREAS, this creature is generally and commonly known as a "Sasquatch", "Yeti","Bigfoot", or "Giant Hairy Ape", and

WHEREAS, publicity attendant upon such real or imagined sightings has resulted in an influx of scientific investigators as well as casual hunters, many armed with lethal weapons, and

WHEREAS, the absence of specific laws covering the taking of specimens encourages laxity in the use of firearms and other deadly devices and poses a clear and present threat to the safety and well-being of persons living or traveling within the boundaries of Skamania County as well as to the creatures themselves,

THEREFORE BE IT RESOLVED that any premeditated, wilful and wanton slaying of any such creature shall be deemed a felony punishable by a fine not to exceed Ten Thousand Dollars ($10,000.00) and/or imprisonment in the county jail for a period not to exceed Five (5) years.

BE IT FURTHER RESOLVED that the situation existing constitutes an emergency and as such this ordinance is effective immediately.

ADOPTED this 1st day of April, 1969.

SEAL — COUNTY OF SKAMANIA, STATE OF WASHINGTON

BOARD OF COMMISSIONERS OF SKAMANIA COUNTY

By Conrad Lundy Jr.
Chairman

APPROVED:

Robert K. Leick

Skamania County Bill 69-01

ORDINANCE NO. 1984-2

PARTIALLY REPEALING AND AMENDING ORDINANCE NO. 1969-01

WHEREAS, evidence continues to accumulate indicating the possible existence within Skamania County of a nocturnal primate mammal variously described as an ape-like creature or a sub-species of Homo Sapiens; and

WHEREAS, legend, purported recent findings, and spoor support this possibility; and

WHEREAS, this creature is generally and commonly known as "Sasquatch", "Yeti", "Bigfoot", or "Giant Hairy Ape", all of which terms may hereinafter be used interchangeably; and

WHEREAS, publicity attendant upon such real or imagined findings and other evidence have resulted in an influx of scientific investigators as well as casual hunters, most of which are armed with lethal weapons; and

WHEREAS, the absence of specific national and state laws restricting the taking of specimens has created a dangerous state of affairs within this county with regard to firearms and other deadly devices used to hunt the Yeti and poses a clear and present danger to the safety and well-being of persons living or traveling within the boundaries of this county as well as to the Giant Hairy Apes themselves; and

WHEREAS, previous County Ordinance No. 1969-01 deemed the slaying of such a creature to be a felony (punishable by 5 years in prison) and may have exceeded the jurisdictional authority of that Board of County Commissioners; now, therefore

BE IT HEREBY ORDAINED BY THE BOARD OF COUNTY COMMISSIONERS OF SKAMANIA COUNTY that that portion of Ordinance No. 1969-01, deeming the slaying of Bigfoot to be a felony and punishable by 5 years in prison, is hereby repealed and in its stead the following sections are enacted:

SECTION 1. Sasquatch Refuge. The Sasquatch, Yeti, Bigfoot, or Giant Hairy Ape are declared to be endangered species of Skamania County and there is hereby created a Sasquatch Refuge, the boundaries of which shall be co-extensive with the boundaries of Skamania County.

Skamania Co Ord 1984-2, page 1

SECTION 2. Crime - Penalty. From and after the passage of this ordinance the premeditated, wilful, or wanton slaying of Sasquatch shall be unlawful and shall be punishable as follows:

(a) If the actor is found to be guilty of such a crime with malice aforethought, such act shall be deemed a Gross Misdemeanor.

(b) If the act is found to be premeditated and wilful or wanton but without malice aforethought, such act shall be deemed a Misdemeanor.

(c) A gross misdemeanor slaying of Sasquatch shall be punishable by 1 year in the county jail and a $1,000.00 fine, or both.

(d) The slaying of Sasquatch which is deemed a misdemeanor shall be punishable by a $500.00 fine and up to 6 months in the county jail, or both.

SECTION 3. Defense. In the prosecution and trial of any accused Sasquatch killer the fact that the actor is suffering from insane delusions, diminished capacity, or that the act was the product of a diseased mind, shall not be a defense.

SECTION 4. Humanoid/Anthropoid. Should the Skamania County Coroner determine any victim/creature to have been humanoid the Prosecuting Attorney shall persue the case under existing laws pertaining to homicide. Should the coroner determine the victim to have been an anthropoid (ape-like creature) the Prosecuting Attorney shall proceed under the terms of this ordinance.

BE IT FURTHER ORDAINED that the situation existing constitutes an emergency and as such this ordinance shall become effective immediately upon its' passage.

REVIEWED this 2nd day of April, 1984, and set for public hearing on the 16th day of April, 1984, at 10:30 o'clock A m.

BOARD OF COUNTY COMMISSIONERS
Skamania County, Washington

[signature] Chairman

[signature] Commissioner

[signature] Commissioner

ATTEST [signature]

County Auditor and Ex-Officio Clerk of the Board

ORDINANCE NO.______
Page 2 of 3 Pages

Skamania CO ord 1984-2, page 2

ORDINANCE NO. 1984-02 IS HEREBY DULY PASSED AND ADOPTED INTO LAW this 16th day of April, 1984.

BOARD OF COUNTY COMMISSIONERS
Skamania County, Washington

G. Callahan
Chairman

Eric Widun
Commissioner

William V. [illegible]
Commissioner

ATTEST:

Gary M. Olson
County Auditor and Ex-Officio Clerk of the Board

ORDINANCE NO. 1984-2
Page 3 of 3 Pages
Pros. Attny.

Skamania Co Ord 1984-2, page 3

bigfoot.res 6/9/91

SPONSORED BY: Consent

PROPOSED BY: Harris

INTRODUCTION DATE: 6/9/92

1 **RESOLUTION NO. 92-043**

2 **DECLARING WHATCOM COUNTY A SASQUATCH PROTECTION**

3 **AND REFUGE AREA**

4 WHEREAS, legend, purported recent findings and spoor suggest that Bigfoot may
5 exist; and

6 WHEREAS, if such a creature exists, it is inadequately protected and in danger of
7 death or injury;

8 NOW, THEREFORE, BE IT RESOLVED by the Whatcom County Council that,
9 Whatcom County is hereby declared a Sasquatch protection and refuge area, and all
10 citizens are asked to recognize said status.

11 BE IT FURTHER RESOLVED, this resolution shall be effective immediately.

12 APPROVED this 9th day of June, 1991.

13 WHATCOM COUNTY COUNCIL
14 ATTEST: WHATCOM COUNTY, WASHINGTON

15 Ramona Reeves [signature] Daniel M. Warner [signature]
16 Ramona Reeves, Council Clerk Daniel M. Warner, Chair

17 APPROVED AS TO FORM:

18 ______________________
19 Civil Deputy Pros. Atty.

Whatcom Co Ord #92-043

Chapter 2
Starting the Search

When I discovered that various levels of government had acknowledged that Bigfoot exists and actually enacted legislation, this was enough to get my motor going and start a serious hunt as to where I'd search. I set out to determine where in California, Oregon or Washington justified an ongoing investigation of these phenomena. After weeks of working the Internet and deciphering graphs and charts it became obvious that there was more than one place to look. When you read the reports from the surrounding mountains of Hoopa, this valley becomes even more interesting. I took a trip to the U.S. Geological Survey, pulled a mountain of topographic maps and spent days looking at the areas that had reported Bigfoot sightings. My next step was contracting for a series of aerial photos and determining where the residences and urban settings were in relationship to sightings.

The Hoopa Valley is one of only two valleys in this region of Northern California, the other being the Scott Valley further east. To the north of Hoopa is the Klamath River where Bluff Creek cuts across a mountain range. This region was made famous because of the Patterson–Gimlin footage in 1967 of Bigfoot running across Bluff Creek. During the late 1950s and early 60s Bluff Creek was a hotspot for Bigfoot sightings, footprints and even some funny hoaxes. There have been many visitors to this region, but few actual sightings in the new century. I was in the middle of this area in the summer of 2006 and camping at Louse Camp, which is located on Bluff Creek approximately six miles south of the location where the Patterson–Gimlin film was taken. I was with two friends when we arrived at the campground. We found a tent, small fire and a very nicely dressed man, too nicely dressed. I engaged the fellow in conversation and found that he was from Boston, flew to San Francisco, hitchhiked all the way to Orleans and got a ride from a U.S.F.S. maintenance crew into Louse Camp. I asked him why he ended up here. He said that he wanted to see Bigfoot so he made the trip. The person had no transportation, no water, and no water filtration and didn't appear to have any of the needed essentials for a trip into the backcountry. I asked him how he was going to get out when he was ready to head home. He told me that the maintenance crew told him they came by every 4–5 days and they would take him out. He was there for two more days, then left in the middle of the day while I was hiking. I would caution anyone who wants to come to this region, don't, unless you are a seasoned camper. The weather can be treacherous, the trail is very tough and you won't see Bigfoot. You had a chance of seeing him in this region in the 1950s and maybe early 60s, but not now.

Hoopa has a valley elevation of 350 feet. According to the 2000 United States Census there were 1,893 Hoopa tribal members living in the valley, 337 from other tribes and 403 non-Indians. The total population of Hoopa in 2000 was 2,633.

The Hoopa people were fortunate in 1876 when President Ulysses Grant granted them the largest reservation in California, 144 square miles (89,572 acres), roughly the size of the city of San Jose (which has close to one million people). The reservation has lush forests, a huge valley, a beautiful river and many creeks flowing through it.

The weather in Hoopa is one feature that offers an interesting mix for Bigfoot hunters.

	Mean Avg/350'	At 2300 Feet	Annual Rainfall
Summer	70.9 Degrees F	63.76 Degrees F	
Winter	45.1 Degrees F	37.97 Degrees F	58.35 Inches

Snowfall on the valley floor is minimal, 0.4 inches annually. The temperature figures at 2,300 feet should be noted and will be addressed in a later chapter where Bigfoot lifestyles will be examined.

Approximately twenty miles to the west of Hoopa is the California coastline, which includes Eureka, Redwood National Park, Arcata, Highway One, Bald Mountains and a series of cities where Bigfoot has routinely made its appearance. The mountain roads leading west from Hoopa through this range are windy and treacherous (except the main highway). Many of the roads carry very little vehicular traffic and have very few homes. The mountains and forests in this area are known for producing California's largest cash crop, marijuana. Hunters, fishermen and hikers must be careful when exploring this range because of the hazards associated with stumbling upon a growing marijuana field. Many of these areas have booby traps, armed guards and nasty owners. I'd love to hear some of their stories about a Bigfoot stumbling into their fields and how they responded, but I don't think they'd have filed a sighting report.

Approximately 12 miles from the northern Hoopa reservation property line is Bluff Creek. The next major creek north of Bluff Creek is Blue Creek. Bluff Creek starts in a valley just west of the Go Road that starts near Orleans. The government started the Go Road in the late 1950s with the idea to make a road all the way through the Siskiyou National Forest and onto Highway 199. The workers were able to cut the road almost 30 miles up from Orleans and across some very interesting terrain. It was during the cutting of the dirt road that stories started to emerge of giant human-like footprints, huge metal barrels thrown over the road, tractor tires weighing 200 pounds moved and more. Road workers started to say that there was a huge creature roaming the area that was destroying their construction equipment. The road was adjacent to Bluff and Blue Creeks and started less than seven miles from the northern border of the Hoopa reservation.

It was the Hoopa and Yurok Tribes that went to federal court and claimed that the government was building a road through sacred territory. The natives stated that just because the area was not a reservation, it did occupy sacred peaks, mountains and valleys, and the road destroyed the serene nature they needed to hold religious ceremonies. After months of legal fighting, a federal judge ordered the government to cease building the Go Road.

It was during the Go Road construction and associated timbering around Bluff and Blue Creeks that stories of Bigfoot continued to proliferate. Some locals refused to work the road and the area because of fears of seeing the creature or its tracks.

It was in October of 1967 when Roger Patterson and Robert Gimlin decided to travel to Bluff Creek and see if they could film a Bigfoot. Bluff Creek is approximately 38 air miles from Oregon, with very few homes between Orleans and Highway 199. Bluff Creek is 19 miles from the Pacific Ocean and close to the Klamath River in which it flows.

On October 20, 1967 Patterson and Gimlin left the area of the Go Road and started to slowly make their way by horseback down into the Bluff Creek region. Just as they were about to reach Bluff Creek they each saw movement in the creek. Patterson's horse started to act up, so he jumped off while holding his 35mm Kodak movie camera. Patterson was able to get some film of a female Bigfoot crossing Bluff Creek. It totaled 24 feet of footage, some jumpy and hard to view and some very good footage of the creature. The Bigfoot was walking on two legs, had very long arms, black hair covering its entire body and face, and large breasts were visible. Experts have reviewed the footage extensively and guess the creature was seven feet, three inches tall and weighed in excess of 700 pounds.

Since the initial footage of Bigfoot was shot, the film has been reviewed and seen by thousands. In the last 10 years experts have digitized and slowed the film for review. The footage now shows that the creature appears to be carrying something in its left hand. It also shows muscles moving in the right thigh as the creature puts weight on that leg and muscles can be seen in the right shoulder area as it moves its arm.

Movie experts from Disney and other top labs at the time stated that they did not have the expertise to build a costume that was able to expose muscle as in the Patterson–Gimlin movie. Professors who are experts on body movement have stated that there is not a human in existence that could move and walk in the manner the creature did on tape. Forty years after the filming of the movie this still stands as the greatest footage in existence of a Bigfoot. At a time of huge technological advancement and public interest, the Patterson–Gimlin footage continues to be an amazing look at a creature that was seen just miles from the Hoopa reservation.

South of Hoopa is the town of Willow Creek. It has made itself the Bigfoot capital of the world and annually holds Bigfoot Days on the Labor Day weekend. This is a huge event that draws university professors, professional Bigfoot hunters and a variety of amateur explorers. It's an interesting event that is worth attending if you are interested in the region and Bigfoot. In an extensive search of sightings and their locations, Willow Creek came up with relatively few compared to other regions in Northern California, however, they do hold a good Labor Day event and they have an interesting Bigfoot museum in town.

To the east of Hoopa are some of the most desolate areas of Northern California. This region is predominantly Hoopa reservation property. The Trinity Alps Wilderness Area is east of the reservation. All vehicles are excluded from this area. Only transportation by horseback is allowed into the wilderness. The area east of Hoopa and inside the boundary of the reservation includes Tish Tang Creek, Mill Creek, Red Cap Creek, Mill Creek

Road and many, many others noteworthy for Bigfoot encounters. The Hoopa Tribal Forestry group maintains this area, some of which is contracted out for specific forestry cutting and logging. There are no public campgrounds anywhere in this region and the public has restricted access to the reservation without a specialized permit. The public can drive on any paved public road, but they are not allowed on any dirt roads in the Hoopa reservation.

The list below is the result of an Internet sightings search in the greater Hoopa area. The databases and websites are listed later in the book. Some of the reports have little information with few names.

1958	Bluff Creek: Multiple sightings by several individuals.
1958	A woman and her daughter see a large and a small Bigfoot on a hillside above Hoopa Valley just northwest of Willow Creek
1960	Leroy Doolittle sees Bigfoot standing in meadow near Hoopa, specific location unknown.
Jan 1960	Red Cap Creek at Willow Creek. Bigfoot bedding found up hillside, six miles from south of creek.
August 4, 1960	Bob Titmus sees two sets of Bigfoot tracks, the same as those he had seen a week prior on Mill Creek Ridge Rd, eight miles south of Hoopa.
July 1963	Peters saw a Bigfoot jump out in front of him and then jump a five-foot fence and leave the area.
August 3, 1963	A man and his son see a Bigfoot leap over a five-foot fence and run into the woods near Hoopa. Exact location of this sighting is unknown.
Sept 1969	Bigfoot sighted at Karok Indian reservation Klamath at Trinity River.
Oct 1974	Fisherman sees Bigfoot holding fish at Klamath and Trinity Rivers.
August 1988	Grandfather and grandson smell Bigfoot and see tracks up Big Hill Road. (Eastern foothills of Hoopa.)
Oct 1988	Bigfoot sighted at Hwy 96 at Hoopa.
July 1989	Tish Tang Campgrounds. Bigfoot tracks seen in sand along Trinity River across from Tish Tang Creek.
1996	Pecwan Creek at Yurok Reservation at at Jump Dance. Trinie Inong and cousin Elizabeth saw a Bigfoot, almost nine feet tall.
May 2001	Bald Hills near Weitchpec. Sounds of screams heard and an animal's red eyes were seen in forest.

This sightings list offers an overall glimpse of the activity in and around Hoopa. There are many more sightings that occurred in the periphery of the valley, but were excluded because they cannot be directly linked to Hoopa. One amazing aspect to each of these sightings is that the Hoopa Valley has been an area with significant human habitation for over 250 years. Bigfoot knows that people have lived here for a long time, and probably knows they won't be leaving anytime soon. It would appear that Bigfoot has become acclimated to people and has learned to live around Hoopa residents. If you study the sightings list, there are a few glaring realities that are easy to decipher. I placed pins on a map where the sightings occurred, and also did this with aerial photos. Bigfoot can be placed in or near a creek or river with every sighting. One sighting specifically — Red Cap Creek — mentioned that bedding material was found on a small hillside just above the creek. I looked further into this sighting and found that it was almost eight miles from its headwaters, and deer bones were found adjacent to the bed. The animal had made the bedding from moss and grass in the area to a depth of 6–8 inches. In one of my many conversations with Tony Hacking (Chief Wildlife Biologist for the U.S.F.S. in Orleans) he told me of another sighting of bedding material in a creek just north of the Klamath River. Tony stated that the Forest Service was conducting a fisheries study and two scientists had gone into that area in the early summer months of 2006. They had found bedding material similar to what I described on Red Cap Creek. The scientists also found a pile of deer bones adjacent to the bed and couldn't figure out what type of animal could make such a bed or leave a pile of bones. Tony was on another fisheries study and found a very large game trail next to Blue Creek. He said that the trail was very well worn, but also very unusual. He said that the trail had very high clearance, 7–8 feet, and was also cleared as though something wide had been walking there, maybe 4–5 feet wide. Tony explained how this area was the next valley north of Bluff Creek from the Klamath and had no roads or fire trails of any kind. Tony said that this valley is so desolate that they had their team park their cars at the top of the valley and they had another team member park their car at the bottom of the valley. They felt they would never be able to get in and out in one day if they had to climb back out the way they went in. Tony said that when he was in the valley he felt it was very wild, and a prime Bigfoot spot.

The decision to work in Hoopa couldn't be finalized until I looked at the overall sighting numbers throughout the western U.S. I know that this sounds like an analytical adventure to pacify my curiosity, but there was a method to the madness. I needed to understand the area where Bigfoot lived, what attracted it to the area and is there a geographical or environmental factor that keeps it returning.

By The Numbers

When trying to understand Bigfoot and deciding where to search I think it's important to understand its geographical distribution and environment. Bigfoot sightings have been reported in every state of the United States except Rhode Island and Hawaii. Since I live in California and I have focused on Northern California I'll explain one rationale of how I ended up in Hoopa.

When I was in college I took a class about the practical applications of math and statistics. In thinking back to those college days I would never have thought I'd be using my knowledge to map out Bigfoot. I don't think it would have been possible to extrapolate the data about Bigfoot when I was in college. The advent of the Internet, online reporting and the assimilation of databases have made tracking information more user friendly.

There are several widely known Bigfoot/Sasquatch databases and books that accumulate sightings and their associated data. I have not found a database that has put this information together and tried to statistically show patterns of activity. The databases and books that I utilized to accumulate sightings include:

Nabigfootsearch.com
CaliforniaBigfootsearch.com
BFRO.net
GCBRO.com
OregonBigfoot.com
Bigfootinfo.org
Home.clara.net
Bigfootencounters.com

I documented all Bigfoot sightings that I could locate in Washington, Oregon and California. The two states with the most reported sightings are California (approximately 400) and Washington (approximately 450). I decided to document all of the sightings in California that were significant in grouped numbers and geography, as this would assist me in understanding Bigfoot distribution in a geographical region. In order to be considered significant, a sighting had to include information other than just "Bigfoot sighted, Highway 37, 9 a.m., 1966." I looked for dates and times, a narrative describing the circumstances, a description of the creature, and possibly the chance that the sighting was investigated to some degree.

It became quickly apparent that large numbers of California Bigfoot sightings were centered in the far northwest corner of the state. The location of the California sightings obviously led to the next question: How many sightings were reported in Oregon? I don't think that Bigfoot looked at the geography of California, saw the Oregon border and decided not to cross, so, it would seem obvious that an adjacent county in Oregon should have similar Bigfoot statistics as its California neighbor.

The counties, Del Norte, Humboldt and Siskiyou were the top three counties in California for the number of reported Bigfoot sightings, and they were also the counties in the farthest northwest corner of the state. As I began to accumulate data for adjacent Oregon counties I started to notice a definitive pattern. The most southern Oregon counties of Curry, Josephine and Jackson all had similar sighting numbers relative to their size. I decided to keep going north to determine how far north the current rate of sightings continued.

The further north I accumulated sightings data into Oregon the more that a distinct pattern emerged. I went all the way to Washington accumulating information and found that the western third of Oregon had most of the sightings for the entire state. I took the eastern edge of the following counties as my stop point:

Hood River
Clackamas

Marion
Linn
Lane
Douglas
Jackson

The western edge of the sighting region was the Pacific Ocean. The eastern edge of the sighting range was the Cascade Mountain range. Since there was a distinctive pattern in Oregon and California, I decided to keep going north.

As I started to accumulate data into Washington the same line of demarcation again was readily apparent. I used the eastern edge of the following counties as my stop point:

Whatcom
Skagit
Snohomish
King
Thurston
Lewis
Skamania

The Pacific Ocean formed the western edge of this sighting area. The eastern edge of the Washington sightings was the eastern slopes of the Cascade Mountain range. I studied elevations, populations, food sources, weather, precipitation and tourism numbers, all in an effort to reconcile the high numbers in the associated counties. It wasn't until I reviewed a weather chart for the Western United States (specifically Oregon) did the sighting numbers make complete sense. [See color version on page 193]

The eastern edge of the dark blue section is also the exact eastern edge of the Oregon counties noted for large numbers of Bigfoot sightings. The western (left) edge of the map is the Pacific Ocean. The northeastern portion of the diagram shows a small green and blue area that huddles near the eastern edge of the Okanogan County. Okanogan County is the one county in that entire eastern region that has a slight blip of Bigfoot activity with 18 reported sightings, obviously in conjunction with the precipitation reading on the map.

Oregon Precipitation Map. [See color version on page 193]
Copyright 2007 PRISM Group, Oregon State University, www.prismclimate.org

All of the 17 counties to the right of the blue area on the map have a combined total of 135 sightings versus the 19 counties in the blue, which have a total of 446 sightings. The 19 counties represented by the blue have one-third of the total landmass of Oregon, yet include over 75 percent of the Bigfoot sightings.

The satellite image on page 38 [See color version on page 194] shows a silver box with California at the bottom, Oregon in the middle and Washington at the top. The state boundaries are outlined in red. The area highlighted in silver shows the region with the highest frequency of Bigfoot activity. This satellite photo was chosen because the cloud lines in Washington almost exactly follow the Bigfoot counties cited in the graphs on pages 40 and 41. The far northwest corner of California, which has Humboldt and Del Norte Counties, has an almost half-moon-size cloud band indicating its boundaries. The map shows that clouds occupy these counties and thus they have high levels of precipitation.

The U.S. map on page 39 [See color version on page 193] eloquently depicts the northwestern portion of California and the western sections of Washington and Oregon that have been documented to have the most

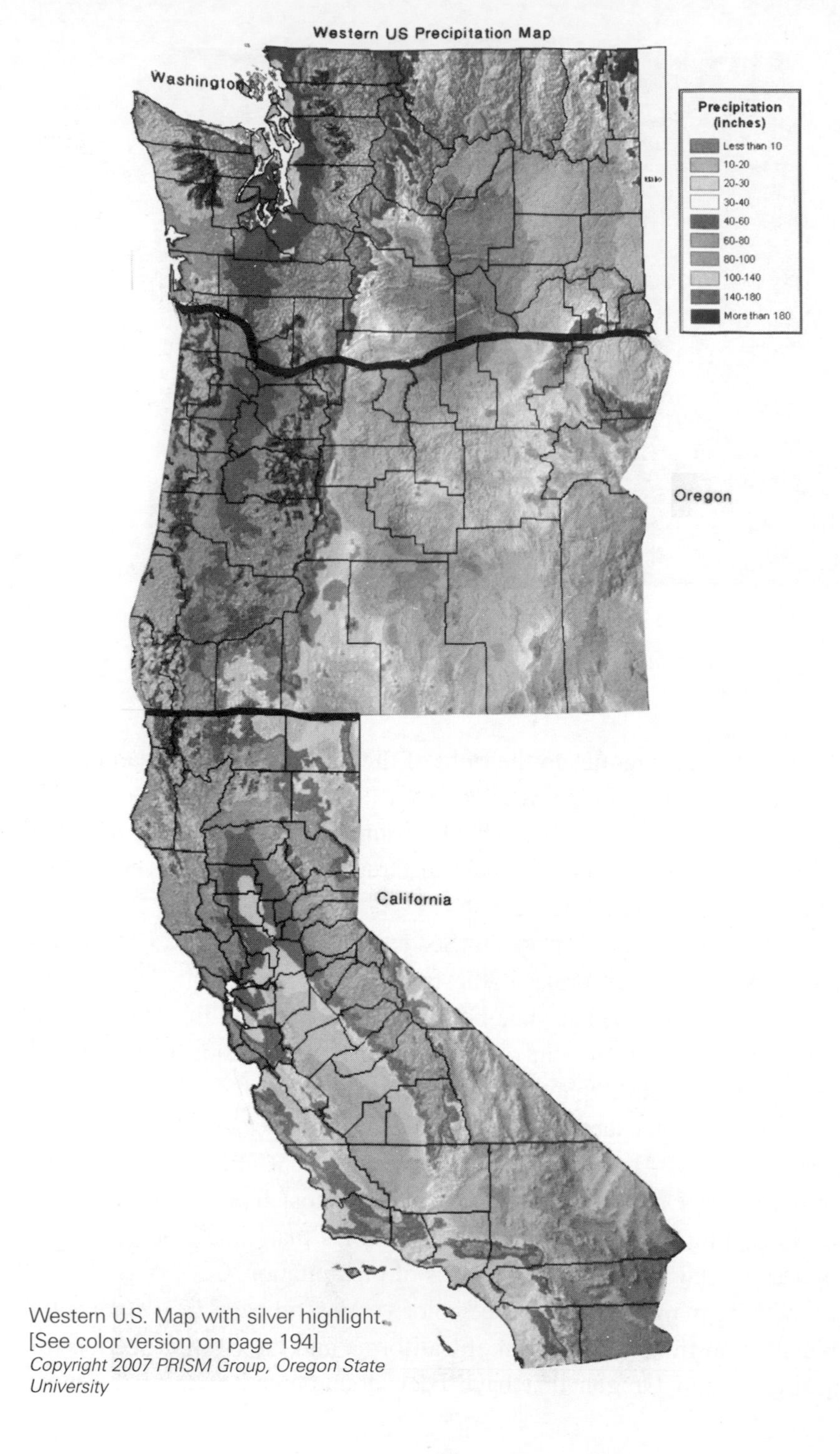

Western U.S. Map with silver highlight.
[See color version on page 194]

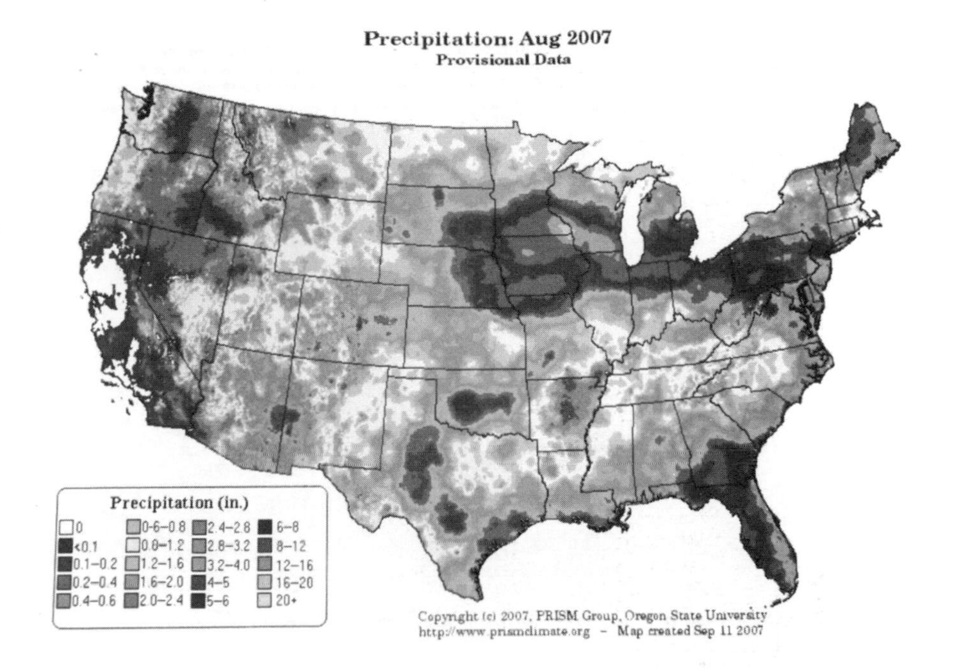

U.S. precipitation map. [See color version on page 193]

Bigfoot sightings. The blue section of those three states perfectly outlines the area identified as having high numbers of Bigfoot sightings. The borders are so well defined that in looking at the Oregon rainfall map it almost exactly follows the borders of those counties. Coincidentally, the one county in all three states that has the most Bigfoot sightings is also one of the counties that has the largest amount of rainfall — Skamania.

While reviewing the United States rainfall map I was also examining the United States distribution of Bigfoot sightings. I do not believe that you can draw conclusions about where Bigfoot activity will be in different parts of the U.S. based on what we know about the Northwest. There is significant Bigfoot activity in Arizona and New Mexico, where there is a dry climate. There have been Bigfoot sightings in the deserts and on the fringes of the deserts of southern California, which obviously show that Bigfoot can adapt if it can survive New Mexico, Arizona and southern California. On the other extreme, there have been Bigfoot sightings at the height of winter in high elevation snow zones, which suggest to some that it may be lost and to others it has adapted to living in higher elevations.

Counties	# of Bigfoot Sightings	# of Years of Sightings	California 2005 County Population	Cty Size in Square Miles	Avg # of Sightings/Yr	Sightings Per cnty sq mile One Sighting Per every...	Sightings Per cnty resident one sighting per every	Rankings- Best Chance to worst to see Bigfoot
Alpine	6	23	1159	738	0.26	124	**193**	1
Butte	19	40	214885	1639	0.48	99	11309	
Calaveras	8	36	46871	1020	0.22	128	5858	
Del Norte	25	26	28705	1007	0.96	**49**	11482	
El Dorado	1	20	176841	1710	0.05	1788	176841	
Fresno	3	6	877584	5962	0.5	2005	292528	
Humboldt	**84**	31	128376	3572	**2.7**	122	1528	6
Inyo	25	29	18156	10203	0.86	162	726	3
Kern	17	31	756825	8140	0.55	480	44519	
Lassen	2	18	34751	4557	0.11	2360	17375	
Marin	2	32	246960	519	0.06	414	123480	
Mendocino	19	43	88161	3508	0.44	204	4640	8
Modoc	3	27	9524	3944	0.11	1401	3174	7
Mono	2	8	12509	3044	0.25	1566	6254	
Monterey	4	20	412104	3321	0.2	942	103026	
Napa	1	28	132764	753	0.04	788	132764	
Nevada	5	50	98394	957	0.1	194	19678	
Plumas	11	**66**	21477	2553	0.17	238	1952	5
Riverside	5	44	1946419	7207	0.11	1460	389283	
Santa Cruz	3	36	249666	445	0.08	202	83222	
Shasta	21	50	179904	3785	0.42	183	8566	
Sierra	3	19	3538	962	0.16	320	1179	4
Siskiyou	30	39	45259	6286	0.77	211	1508	5
Sonoma	6	26	466477	1575	0.23	294	77746	
Tehama	8	42	61197	2950	0.19	370	7649	
Trinity	21	46	13622	3178	0.46	152	648	2
Tulare	2	8	410874	4823	0.25	2419	205437	
Tuolumne	7	43	59380	2235	0.16	324	8482	
Yolo	1	15	184932	1013	0.01	1023	184932	
Totals	**262**	**avg-41**	**6927314**	**91606**	**avg-.38**	**avg-690**	**Avg-26440**	

California Bigfoot Spreadsheet.

It should be obvious at this point that the Bigfoot of the western United States prefers a moist, wet climate. I've included a spreadsheet of California counties reporting Bigfoot activity to outline that Bigfoot is in almost every geographic climate of the state, and the same can be said for Washington and Oregon.

All Bigfoot sightings (area represented in all three graphs) only go back to 1940. Sightings and incidents prior to this date are predominantly secondhand. In attempting to stay with only firsthand accounts of Bigfoot sightings, I used 1940 as the cutoff date. There are several criteria that could be used to measure which county had the highest incidence of Bigfoot activity; the most obvious factor would be the number of Bigfoot sightings. In a county large in landmass the number of sightings may not be an adequate indicator of relative Bigfoot activity compared to a county that was quite small. Because of this concern I developed a category "Sightings per county square mile," meaning that this category represents one sighting per the number of square miles listed in the box. The higher this number equals a higher incidence of Bigfoot sightings.

If a county was very large in landmass and had few residents, there

Washington

Counties	# of Bigfoot Sightings	# of Years of Sightings	2005 County Population	Cty Size in Square Miles	Avg # of Sightings/Yr	Sightings Per cnty sq mile One Sighting Per every...	Sightings Per cnty resident one sighting per every	Rankings- Best Chance to worst to see Bigfoot
whatcom	17	40	183471	2119	0.43	124	**10792**	
Skagit	6	25	113171	1735	0.24	289	18861	
Clallam	11	58	69689	1739	0.19	158	6335	
Snohomish	36	49	655944	2089	0.73	58	18220	
Jefferson	18	26	28266	1814	0.69	100	1570	3
King	35	37	1793583	2126	0.95	61	51245	
Grays	31	35	70900	1916	0.89	62	2287	4
Mason	17	47	54359	691	0.36	**36**	3197	5
Thurston	13	40	228867	727	0.33	56	17605	
Pacific	2	21	21579	932	0.05	466	10789	
Lewis	20	39	72449	2407	0.51	120	3622	6
Wahkiakum	3	28	3849	264	0.11	88	1283	2
Cowlitz	12	55	97325	1138	0.22	95	8110	
Clark	8	23	403766	628	0.35	79	50472	
Skamania	**57**	57	10664	1657	**1**	**29**	**187**	**1**
Pierce	**46**	40	753787	1678	**1.15**	**36**	16386	
Totals	**332**	**avg-39**	**4561669**	**23660**	**avg-.51**	**avg-116**	**avg-13810**	

Rankings= A #1 ranking indicates it is the best chance to view a Bigfoot if you area resident of the county.

Sightings per county resident indicates the number of sightings per the number indicated. #1 ranking=Skamania

Sightings per county square mile divides the square miles in the county by the number of sightings.

County size and county population is taken from the U.S. Census 2005 projections.

Bigfoot Sightings numbers were taken from the California Bigfoot Search database, BFRO database, GCBRO, The Sasquatch File, The Track Record, Sasquatch Information Society, Bigfoot Casebook and Oregon Bigfoot

Washington Bigfoot Spreadsheet.

may not be many Bigfoot sightings. It was important to represent this concern by developing another category on the graph, "Sightings per county resident," meaning this category represents one Bigfoot sighting per the number of county residents listed in the box. The lower the number of residents, the greater the chance of a sighting.

The category "# of years of sightings" indicates the time span that Bigfoot sightings were found for that county.

The chance of a Bigfoot sighting is based on a multitude of factors. I've interviewed many witnesses who were tourists traveling through a county and witnessed the creature crossing a road. The number of tourists a county may have during any given year will directly affect the number of sightings. Since tourists that visit rural campgrounds don't register or pay a fee and often sleep in the back of their car, an attempt to reconcile these numbers would be futile.

The counties that have the highest populations in California are not represented on the graph, as they generally do not have significant Bigfoot activity. That is not to say that Los Angeles and San Diego Counties do not have activity, they do. However, the numbers of sightings

Oregon

Counties	# of Bigfoot Sightings	# of Years of Sightings	2005 County Population	Cty Size in Square Miles	Avg # of Sightings/Yr	Sightings Per cnty sq mile One Sighting Per every...	Sightings Per cnty resident one sighting per every	Rankings-Best Chance to worst to see Bigfoot
Clatsop	17	39	36798	827	0.44	49	2164	**3**
Columbia	6	30	43560	657	0.2	110	7260	
Multnomah	11	47	672906	465	0.23	42	61173	
Tillamook	11	38	25277	1125	0.29	102	2297	4
Washington	14	38	500585	727	0.37	52	35756	
Yamhill	11	23	88150	718	0.48	65	8013	
Clackamas	**118**	46	338391	1868	**2.57**	**16**	2867	5
Hood River	25	39	21284	522	0.64	**21**	**851**	1
Marion	25	36	305265	1183	0.69	47	12210	
Polk	12	16	70295	741	0.75	62	5857	
Lincoln	7	14	45994	979	0.5	140	6570	
Benton	7	18	78640	676	0.39	97	11234	
Linn	22	46	108914	2292	0.49	104	4950	
Lane	57	40	335180	4554	1.43	80	5880	
Douglas	35	51	104202	5036	0.69	144	2977	
Coos	15	58	64711	1600	0.26	107	4314	
Curry	17	38	22247	1627	0.45	96	1308	2
Josephine	28	39	80761	1639	0.72	59	2884	6
Jackson	20	28	195322	2785	0.71	139	9766	
Total	**458**	**avg-36**	**3138482**	**30021**	**avg-.65**	**avg-81**	**avg-9912**	

Sightings prior to 1940 are not be included in the sightings list.

All data regarding population and county size is taken directly from United States Census reports.

Oregon Bigfoot Spreadsheet.

versus the population would skew the numbers so much that they were purposely left off the graph. Bigfoot obviously couldn't live at Market and Van Ness in San Francisco, thus San Francisco County was left off the graph. A majority of the counties with the greatest landmass and where the largest population masses in California reside do not have the habitat on their asphalt streets for a thriving family of Bigfoot, thus, they are not listed.

As I decided to look into Oregon Bigfoot sightings I started my accumulation of data from the same websites as for California. The job was long and cumbersome as I read each report and ensured there was no duplication of sightings. Several of the databases do list reports from other Internet sources, so this task does take a significant amount of time. I also reviewed an Oregon County map and attempted to understand the distribution of Bigfoot sightings. It doesn't take a statistician to see that most of the Bigfoot sightings are on the western side of the eastern slope of the Cascade Mountains. In fact, 75 percent of all of the Oregon sightings are located in just one-third of the state's landmass. The counties having the greatest number of sightings are listed on the Oregon spread-

Counties w/ Most Bigfoot Sightings
Western United States

Counties	# of Bigfoot Sightings	# of Years of Sightings	2005 County Population	Cty Size in Square Miles	Avg # of Sightings/Yr	Sightings Pe cnty sq mile One Sighting Per every...	Sightings Pe cnty resident one sighting per every	Rankings- Best Chance to worst to see Bigfoot
California								
Del Norte	25	26	28705	1007	0.96	49	11482	3
Humboldt	84	31	128376	3572	2.7	122	1528	1
Siskiyou	30	39	45259	6286	0.77	211	1508	2
Totals	139	avg- 32	202340	10865	avg- 1.48	avg- 127	avg- 4839	
Oregon								
Hood River	25	39	21284	522	0.64	21	851	1
Curry	17	38	22247	1627	0.45	96	1308	2
Clatsop	17	39	36798	827	0.44	49	2164	3
Totals	59	avg- 39	80329	2976	avg- .51	avg- 55	avg- 1441	
Washington								
Skamania	57	57	10664	1657	1	29	187	1
Wahkiakum	3	28	3849	264	0.11	88	1283	2
Jefferson	18	26	28266	1814	0.69	100	1570	3
Totals	78	avg- 37	42779	1245	0.6	avg- 72	avg- 1013	
Western United States Top 2 Counties for Bigfoot Sightings								
Skamania	57	57	10664	1657	1	29	187	**
Humboldt	84	31	128376	3572	2.7	122	1528	****

** Rating based upon number of sightings per county resident

**** Rating based upon the number of sightings reported per year

Counties with Most Bigfoot Sightings Spreadsheet.

sheet and I have not listed the remainder of the counties in the state. The state does have a huge lump of data in relation to Clackamas County at 118 sightings. A majority of these reports were listed on the Oregon Bigfoot webpage. The number two county in the state for total sightings, Lane County, has less than half the totals as Clackamas. The smallest county in Oregon in landmass that I have listed, Hood River, will probably give its residents the best chance at a sighting. Hood River is located at the far northern potion of the state with its northern edge on the Columbia River and, incidentally, just downriver from the county in Washington with a significant number of sightings, Skamania.

My research into Bigfoot sightings continued to carry me north and into the State of Washington. The information that I accumulated about Washington Counties validated the hypothesis I had made from the data in California and Oregon. The sightings were greatest west of the eastern slopes of the Cascade Mountains and continued to the Pacific Ocean. The counties east of the Cascades had a significant decline in sightings in comparison to the western section, and five counties in the east had no sightings reported. Skamania and Pierce counties have the highest number of reported sightings and are only separated by Lewis County. Both Skamania and Pierce sit almost on the same longitude lines. Skamania's "sightings per county resident number" is easily the lowest number among all three states. Alpine County in California has a 193 (sightings per county resident), but that is somewhat skewed by only six sightings in 23 years. Trinity County in California sits adjacent to Humboldt County and comes in with 648 sightings with 21 sightings in 46 years.

The last graph is an accumulation of the top three counties for Bigfoot sightings in Oregon, Washington and California. I based the rankings on the numbers of sightings per county resident and on the numbers of sightings per year. It's important to note that almost all of the Internet databases that have sightings listed bulk all Bigfoot sightings and incidents into one category. A Bigfoot incident in this book (for affidavit reasons) constitutes an occurrence that can be directly related to accepted and known Bigfoot behavior. A Bigfoot sighting is when a Bigfoot is actually seen. The databases sometimes differentiate these (as I have attempted to accomplish), but this is not always the case. For the purposes of these graphs, all sightings and incidents are listed as "sightings" and they are bulked together.

Skamania and Humboldt counties definitely appear to be the Bigfoot hotspots of the Northwest. There are some definitive similarities between the two counties.

Comparing Skamania and Humboldt Counties

- Both counties border large bodies of water: Skamania—Columbia River, Humboldt—Pacific Ocean and six separate large rivers.

- Both counties have a great variance in rainfall, from 30 inches in some areas to over 90 inches in others.

- The government is the biggest employer in each county.

- Both counties had a flourishing timber industry in years past and now they are struggling.

- Portions of each county sit at sea level, Humboldt bordering the Pacific Ocean and Skamania bordering the Columbia River, which is essentially at sea level.

- Both counties have gently sloping hillsides, steep and rugged mountains that receive snowfall annually and they have regions in their counties that have been heavily forested.

- Both counties have remote areas that are rarely visited, are rugged and dangerous to travel.

The Decision

After accumulating all of the listed data I had to make a decision about whether to dedicate time to one specific region. Humboldt County jumped out at me as a strong location to start Bigfoot research. Hoopa seemed to be a natural location to set up my office and hang a shingle as it was set in a valley with all of the amenities of home. There was a base of population that had a heritage linked to the area for almost two hundred years. The climate of the Hoopa valley matched the areas outlined as having a high incidence of Bigfoot activity. There was a large body of water associated with the valley, the Trinity River. The area also comes with an interesting history, a reservation, and a region that is almost completely surrounded by wilderness areas (Trinity Alps and Marble Mountain), national park (Redwood National Park) and United States Forest Service Property. The rainfall, climate and characteristics of the surrounding environment make Hoopa a great opportunity for spending a significant amount of time understanding and researching Bigfoot. Is it a coincidence that the purported Bigfoot capital of the world, Willow Creek, is 14 miles south of Hoopa? It is also a bonus that the most famous film footage in the world of a Bigfoot (Patterson–Gimlin) was shot just miles north of the reservation boundary at Bluff Creek.

Hoopa Valley

I have now come full circle and am back at Hoopa again. Here are the specifics and background on the reservation.

Seventy miles east of Eureka is Hoopa, California. The Hoopa Valley is the second largest valley in northwest California. Traveling east from the coastal town of Eureka, you meander through the rolling mountains and thick forests and drive into the town of Willow Creek, which claims to be the Bigfoot capital of the world. Turning north from Willow Creek and following the Trinity River you take a quick 20-minute ride on California Highway 96 and arrive in Takimildin, the Hoopa reservation town. There are several signs stating that you are entering the Hoopa community.

In 1864 the United States government established the Hoopa Valley Indian Tribe reservation. In the 1800s there were approximately 1,000 Hoopa people living in the valley; as of 2000, there were 2,500 native Hoopa people living in the community. The lifeblood of the valley is the Trinity River and the surrounding forests. The river supplies fresh water and a constant changing mix of migrating salmon, steelhead, trout and sturgeon. The rights that the government has given the Hoopa allow them to net fish in the river, something non-Native residents cannot do. The Native people utilize the netting of fish as a staple of their diet and a continuation of their cultural heritage. The people also have the right to hunt at anytime in their reservation, something non-Natives cannot ever do and are restricted in public zones by seasons managed by the State of California Department of Fish and Game. I should state here that California's Fish and Game Department is woefully under-funded. The sales of hunting and fishing licenses help fund the department, but there are too few game wardens to adequately protect the resources. In my thousands of hours in the woods of California I have never seen a warden off a main highway. Again, this is not their fault; they need more wardens.

The spelling and pronunciation of the name are slightly different depending on certain factors. When you talk about the tribe and valley it is spelled "Hoopa." When you refer to the people and their language it is "Hupa."

The forests surrounding the valley are extremely thick with vegeta-

Hoopa Valley.

tion and rich in value. The total size of the reservation is approximately 144 square miles and over 75 percent of the area is designated as a commercial timber district. One-third of all the forests in the reservation have been set aside by the tribal government and cannot be touched for any reason. The main revenue generator for the tribe is harvesting their rich forests. Almost all of the timbering is contracted through external sources, but the Hoopa Tribal Council and Forestry do the maintenance, planting and management of the forests.

The eight miles of the Trinity River that flows through the Hoopa valley attracts a wide variety of recreation and general tourism. Depending on the flow in the river, you will find people jet boating, rafting, fishing, swimming, netting and sunning along the miles of beautiful beaches that exist between Willow Creek and Weitchpec. The Hoopa reservation boundary ends in the north where the Trinity River empties into the Klamath River, where the Yurok tribe claims its territory. The Klamath is the largest river in Northern California and attracts more tourists and press than the Trinity. In recent times there have been lawsuits and newspaper articles about the low water flow of the Klamath and how that has limited the number of fish in the river. The flow issue is related to water being diverted for farming rather than allowing the river to flow high and sustain a greater fish population. Many of the local tribes have been very vocal on this issue.

The Sacred High Country

The northern border of the Hoopa reservation is located adjacent to the Klamath River and near an area that local tribes call the "Sacred High Country." Three different tribes claim the High Country as an extremely sacred region and a place they have frequented for hundreds of years. The area is easy to locate and has beautiful peaks and valleys. Crossing the Klamath River and continuing north through the mountains you come to an area of peaks with beautiful exposed rocks, with names such as Turtle Rock, Chimney Peak and Doctor Rock. With a little imagination you can see how they were named and why they were sacred. The area of the peaks has little tourist traffic and is maintained by the United States Forest Service, and by coincidence is directly in the middle of the Bluff Creek region, the location of the Patterson–Gimlin Bigfoot film.

Turtle Rock in the Sacred High Country and a view looking down towards Blue and Bluff Creeks.

The tribes trek to the Sacred High Country annually to train medicine men, practice their religious dances and pray to their gods. This region has been described in courtroom testimony regarding the protection of the region and the stoppage of the Go Road as the "center of the Indians' universe." Even though the region is not under the control of any one tribe, they have all gone to extreme lengths to preserve the region and its cultural value. The United States Forest Service controls the access to, and maintenance for, the area.

As the tribal elders make their trek from their tribes to the Sacred High Country they have to cross Bluff Creek, one of the most famous creeks in Bigfoot lore. Bluff Creek starts high in the region and meanders through the mountains in a huge semi-circle, eventually making its way into the Klamath River. The creek is a breeding ground for salmon and steelhead in the Klamath. It is an area with very limited foot traffic and is restricted by the state as a no-fishing zone.

The building of the Go Road (now known as the Eyesee Road) didn't proceed as smoothly as many would have wished. As crews were making their way through the mountains they encountered some interesting hindrances. Giant tractors, tires, trailers and the usual construction equipment were on site as they plodded their way through the mountains. Crews would work during the day and then retreat down the mountain at night and stay at local lodges. When they returned to their equipment the next morning they found 500-pound tires thrown over the side of cliffs, tanks overturned, etc. The vandals left behind tracks, giant tracks. Some of the tracks were between 16 and 24 inches long and looked similar to a human's bare foot. The workers also heard strange yells and screams when they worked. The construction crews soon started to talk about giant hairy creatures that lived in the region where they were building the roads. The creatures appeared to be aggressive towards objects, but not people, almost as though they were trying to send a message to stay away from their habitat. Stories of these giant creatures and footprints made their way across the nation and were soon attracting the attention of professional trackers. Patterson and Gimlin happened onto the scene and eventually filmed the Bigfoot in Bluff Creek.

During the last 40 years, people have claimed that the film was a hoax. Some have claimed that they were the one dressed in an ape costume and were filmed by Patterson/Gimlin. Almost every imaginable story has been associated with that footage. However, the film stands as credible and is the only close range extended film footage of the creature in existence.

Since 1967 the allure of Bigfoot has continued. The inexpensive camcorder has made it possible for the common park visitor to catch animals in their natural habitat when 40 years ago the only way to observe these creatures was on a National Geographic special. The technological revolution of the 1980s has assisted in documenting more sightings of Bigfoot, yet even with the advent of the camcorder, it has proven to be an elusive creature. Most sightings of a Bigfoot last less than five seconds, and it usually takes 10–20 seconds to activate most camcorders, assuming it is already in your hand.

Best Bigfoot Evidence, Past & Present

There are many stories related to Bigfoot that go back 200 years. There are documented newspaper articles in the 1800s talking about hairy wild men that ran wild in central California, Washington and British Columbia. There are no verified documented cases of a Bigfoot purposely injuring anyone, anywhere at anytime — an unusual aspect considering it is an immense wild animal.

Newspaper articles do not constitute evidence, but they do document sightings prior to telephones, mass media, jet travel and the Internet. People were reporting on Bigfoot encounters before there was mass reporting of the phenomena and before "Bigfoot" was an accepted name for the creature.

Hair samples from Bigfoot have been recovered and many have been DNA classified. Many of these hairs have come back "not on file" with "primate" being the closest known animal. DNA is the best classification tool known to science, but there is not one centralized database that keeps all Bigfoot DNA. This entire section of Bigfoot investigations is very fragmented, with organizations reluctant to share with others. Anyone collecting hair samples should understand that DNA is only recoverable in the follicle of the hair and it deteriorates rapidly after being released from the animal. The best method to collect and retain hair samples is to put them immediately in medical grade alcohol and get them to a lab as soon as possible, which usually isn't accomplished.

Bigfoot scat has also been recovered and DNA classified. The same results have happened with scat as with hair. The one very interesting aspect is that parasites not known in North America have been found in the scat. Bigfoot scat has been reported to look like a giant human scat, very large.

Footprints from Bigfoot have been reported for decades. Many people who have not read recent articles about footprints immediately will call all of them a hoax and unreliable — far from the truth. There have been several university professors and police technicians who have recently gone public, stating that many of the footprint casts are authentic. They can see dermal ridges in the print (essentially the fingerprint of the foot), which would be next to impossible to hoax. Several of the prints have been examined by anatomy experts who state that the contour, bend and structure of the print could only be made by a primate, and could not be hoaxed without advanced knowledge, which isn't readily available to non-experts. Handprints fall into the same category as footprints, and they have had the same level of success in authentication.

In the last several years a team of Bigfoot researchers used bait to bring a Bigfoot into an area. They deliberately picked an area that was conducive to making prints, and then left food for the creature to eat. When they came back into the area they found that the creature had been there, ate the food and actually sat down and possibly leaned on an elbow while eating its meal. The area where the creature sat was made into a large cast (The Skookum Cast) and has been examined by some of the most advanced scientists interested in Bigfoot in North America. This cast has validated much of the information about the physical aspects (size, weight, body structure, etc.) of Bigfoot.

I am purposely avoiding an extensive description of the Bigfoot evidence because that is not what this book is about. There are many outstanding books in print that offer exhaustive insights into evidence that will satisfy anyone's craving in that area.

The First Contacts

Officer Greg O'Rourke
Officer Chance Carpenter
Hoopa Tribal Police Officers

On my first night in Hoopa I found two tribal police officers parked next to the only gas station in Hoopa. They were talking to each other in separate police vehicles. I drove up in my vehicle, parked and walked up to an officer later identified as O'Rourke. I explained why I was in town (looking for people who have witnessed Bigfoot) and asked if he could shed any light on the issue.

O'Rourke was raised in the Hoopa area and is a Native Indian. He stated that the Hoopa people believe in Bigfoot and that it lives in the trees, "similar to what we saw in the movie Predator." He said that they believe that it lives in the middle zone between both dimensions and can travel between them. He stated that they believe that Bigfoot should be left alone; don't bother it, and it won't bother you.

O'Rourke stated that 30 years ago there was a story that a woman who lived in the mountains between Hoopa and Willow Creek and had found and kept a Bigfoot male child. The story goes that she raised it into its teens; it never went to school and just stayed at the house. He stated that it had hair over its entire body and never spoke to anyone. He said that when the woman died, the male creature escaped into the hills and was never seen again.

O'Rourke's only incident involving Bigfoot was when he was 12 years old. He said he was hunting in the area with other family members when they were setting up camp. They forgot some items in their car and he volunteered to go back alone and retrieve them, carrying his rifle with him. He said that as he reached the car, he heard something strange and he froze. He notes that Indians are always taught that when you hear anything strange or see anything strange, you should immediately freeze. He heard a loud and solid "thump, thump" approximately 30 feet from where he was standing. He maintains it was so solid that the earth almost shook. The footsteps stopped and he felt as though something or someone was watching him. He grabbed what he needed and made a dash back to camp, thinking that maybe one of his family members was playing a trick on him. He stated when he finally got back to his group he found them all there and there was no prank.

Chance Carpenter was the other officer I met on my first day in town. He also grew up in Hoopa and has stories about Bigfoot. He said that he has talked to many people over the years who have seen Bigfoot on the roads leaving Hoopa, usually in the late afternoon and evening. He related how a friend had told him that he was driving up a local road, went around a turn and saw Bigfoot kneeling in the middle of the road. Chance told me that I should meet Inker McCovey at the recreation department as he had a few great stories about Bigfoot.

A story closer to home for Chance is one that involves his friend's girlfriend, who lives on the fringe of town. He said that she heard a noise at the side of her house near her garbage cans and went to investigate. She saw Bigfoot reaching over her shed with two huge arms pulling the bags of garbage over the fence. When she yelled, Bigfoot turned and fled. He said that there were dirt handprints on the bags that were shown to neighbors. Chance told me that he would contact this individual and see if she would talk to me. I was later put in contact with witness Raven Ullibarri (see her sighting report for details) who was the woman Officer Carpenter identified as seeing the Bigfoot at her shed.

Klamath River Monster/Kamoss

On a side note, both officers asked if I was interested in hearing a story about a monster that lives in the Klamath River. They state that it looks like an extremely large snake and has the ability to go to the ocean in low water flow. They claim it also has the ability to live in sewage systems near cities that border the ocean. They state this is a very old Indian story that has been validated by many who live in the area, and that many people in the community have seen and experienced the monster, which they called Kamoss. They said that all tribal members tell their children not to enter the Trinity River unless you can see the bottom; if you can't, then Kamoss may be looming in the river. If the river is clear, then it is most likely a low and lazy flow and safe for swimming.

Tony Hacking
Staff Biologist
United States Forest Service
Orleans

On my second day in the Hoopa area I decided to meet with the staff biologist for the Bluff Creek region surrounding Hoopa and Six Rivers National Forest Region, Tony Hacking. I met with Tony at his business Office in Orleans, California. Tony told me that he has been assigned as the "go-to" person on all issues regarding Bigfoot. He stated that his supervisor is the District Ranger and he believes that as long as it brings recreation and interest to the district, he is "for him" spending time on the subject.

I should state here that this is an unusual position for a United States Forest Service Ranger to take. Most USFS supervisors refuse to talk about Bigfoot and won't allow their subordinates to talk about anything related to Bigfoot while on government time.

Tony stated he was recently highlighted on a Travel Channel special about Bigfoot and was asked to give his "expert" opinion and enlighten America on the topic. He took the TV crew to the original site for the famous Patterson–Gimlin Bigfoot video that was filmed in the Bluff Creek Basin. He said that it was quite a challenge first to find the site, and then to positively identify it was even more of a challenge. He said that they confirmed the sighting by finding a pyramid shaped rock that was in the original clip, but that the entire basin has changed dramatically. Tony explained that since 1967 there has been significant erosion and damage to the creek bed in this area, and it looked quite different than it did in the film.

Tony explains that he wants to be a Bigfoot believer and that his position is enthusiastically optimistic. He stated he knows about the dermal ridges in the foot castings, the unknown determination of DNA analysis and the enormous number of stories that proliferate throughout the region. He said that he is the only government biologist anywhere in the basin, and he is covering two different offices. He said that there is supposed to be a position in Happy Camp, but the Forest Service has not filled it.

Hacking said that many people who see Bigfoot are reluctant to talk about the experience for fear of ridicule. Local residents usually direct witnesses to his location for documentation and wildlife identification. He does state that he tries to find a rational explanation for a sighting that is reported, but sometimes it cannot be explained. He said that administrators inside the United States Forest Service (USFS) have refused to expend research dollars on the Bigfoot issue. He explains that his position doesn't do research, it's more about wildlife management. The researchers are in Eureka and they don't, or won't, work Bigfoot-related issues.

Tony Hacking in front of his U.S. Forest Service office.

Tony's job requires that he be in the field a majority of his time doing studies on a variety of wildlife management problems. He said that whenever he is out in the field, he is always looking for evidence of Bigfoot. One piece of evidence that people don't usually search for is wildlife paths that have a height clearing of over seven feet. He points out that most animals that walk trails need a maximum height of five feet to cleanly walk day in and day out. He was recently working in the Blue Creek Trail/Drainage area and walking a wildlife path when he realized he was walking in heavy foliage that had over a seven-foot height clearing. He found that highly unusual for a very desolate area of his region. Tony claimed this is the one time in his career that he felt that he might be walking on a Bigfoot path.

In the same general area of the main section of Blue Creek, a USFS employee working on a special project for the Yurok Fisheries came into his office claiming he found a Bigfoot bed. He told them that they were in the region doing some special environmental studies when he was following the main segment of Blue Creek. Just up one small hill from the creek he found bedding material (straw, dried grass, leaves, etc) that had been piled together in a deliberate attempt to make a bed. He also stated that there were bones lying on the ground immediately adjacent to the bed. He took a photo of the accumulation and showed it to Tony. Tony told me that the photo was a little too grainy for positive identification, but that it was always a possibility Bigfoot was involved.

The statement regarding the bedding does fall into a category of sightings made by other individuals about possible Bigfoot sleeping areas. Many times these beds are directly over/above a creek/river and are often associated with dry foliage, and bones are sometimes found in the area. Tony did point out that some cougars have similar bedding to raise and harbor their young. The site that the USFS employee found was much larger than a normal cougar/mountain lion bed.

I asked Tony to give a best guess of where he would set up a study for the environmental and physical study of Bigfoot. He said that he would drive to the end of the Go Road. (EyeSee Road), noting that it takes approximately one hour to drive all the way to the end. He said that if you go to the northeast, it drops into the Dillon Creek Basin, and the other direction goes into Fish Lake and Bluff Creek area. Both areas are extremely desolate with few people traveling into the region. He said that Upper Blue Creek near the Bluff Creek region would be an excellent choice.

Near the end of our meeting Tony told me of a phone call he received after the airing of the Bigfoot special on the Travel Channel. It was from a female archaeologist from the USFS, Plumas Division. She had discovered a petroglyph painted by Indians, which appeared to be several hundred years old. She had come out and publicly stated that one of the drawings was of a Bigfoot. She was ridiculed, and restricted by her supervisors from ever saying anything publicly about what she had found. She was surprised that Tony's supervisors had allowed him to be shown publicly on television in his uniform speaking about the Bigfoot phenomena. Tony said that he felt fortunate to have the subtle support from his supervisors on the Bigfoot issue.

I invited Tony to join me at a local restaurant for lunch, and he obliged. We spent approximately four hours talking about Bigfoot. He invited me back for a second meeting the following morning at ten o'clock at his office. He stated that he would go over maps of the surrounding area and further explain sighting locations.

The next day Tony supplied me with the needed information to physically inspect several areas of interest. Sadly, Tony passed away from cancer in early 2007 leaving behind a wife and children.

Karuk Tribal Offices

I drove to the tribal offices of the Karuk tribe, which was located in Orleans. I went to their Department of Natural Resources where I met with two tribal members working in the office. I explained why I was in his area and asked if they would talk to me about their tribe and local Bigfoot issues. They stated that they have lived in the area their entire lives and they had no firsthand accounts of Bigfoot. They did, however, say that they would talk, if they could, about their tribal customs. They invited me to ask questions.

I asked them both to explain who the "little people" were. They looked at each other with stern expressions and said that were told by tribal elders that they were not to talk about this. I explained that I did not want them to betray their tribal beliefs or customs or to compromise their personal religious beliefs, but that I knew many people were very interested. They thanked me and said they could talk minimally about it. I expressed my gratitude and said that if I asked an inappropriate question, I apologized in advance and asked for their forgiveness. I did ask them why there was a veil of secrecy about the little people. Both stated all of their people came from the little people. That is where their religious beliefs come from and where their genetic start was located. They both stated that theirs was the first tribe in the region.

Both confirmed that medicine people inside the tribe still have contact with the little people. There are nearby mystical mountains and caves that are sacred and these are the locations where the little people live and where they came from. These locations are hidden and secret. Both agreed that they could not state where these mountains are located. I asked them if there is a written doctrine/bible for their tribe, and they said that there was not. They state that their tribe controls 1.4 million acres and that they believe that their god gave them hands to help nature. They believe they were the first people in the United States. They stated that the main source of income for the tribe was grants from the U.S. government.

After approximately 30 minutes of discussion, I thanked both men for their time and told them that I truly appreciated their outlook and explanation of their tribal beliefs.

After many months of conducting research and interviewing tribal members, it appeared that the sacred area that the Karuk members discussed is the region at the end of the Go Road, the Sacred High Country.

Deputy Greg Berry
Resident Deputy, Orleans
Humboldt County Sheriff's Department

As I was driving back through Orleans I saw a Humboldt County Sheriff's vehicle stopped by the side of the road. I pulled over and engaged the deputy in a conversation. I explained why I was in his region and asked if I could ask him a few questions. He cheerfully said, "Sure."

It was an extreme pleasure to meet Deputy Berry. He had been the resident deputy for Orleans for the past eight years. He knew everyone in the community and had a wealth of knowledge about Bigfoot. He told me that he was married to a tribal woman. Her mother claimed to have spoken to Bigfoot in her backyard at the base of a mountain on several occasions. Everyone in the family believed her, he said, but unfortunately she had passed on. I told him about the people I had met with earlier in the day and Greg assured me that these were great contacts with credible people.

I asked the deputy how isolated the Go Road and the region of Blue Creek and Dillon Creek really were. He asked if I had some time to take a ride with him in his sheriff's vehicle. I said, "Absolutely," and Greg took me to the top of Go Road. During the entire journey we talked about his agency, Bigfoot contacts, tribal issues and the challenges of being a deputy in a small rural community. It was an absolutely fascinating insight into a job that few people truly understand, and I came to appreciate the role of law enforcement in that very rural environment.

We reached the end of Go Road and the view was unbelievable. You could see the entire Bigfoot range, Blue Creek Basin, Dillon Creek drainage and outwards towards Happy Camp. It was a clear day in late June and you could still see the snowy caps of the Marble Mountains.

A couple of interesting events occurred on the ride down the mountain. Immediately after making the turn at the end of the road, we saw a very large pile of scat, probably 1–2 days old, in the middle of the road. The scat was almost two inches in diameter, round and full of stringy foliage. Greg stated he thought it might be a large elk's, but wasn't positive. It was large!

Approximately halfway down the road, we turned a curve and saw a huge black bear meandering down the side of the shoulder of our road. We guessed the bear was between 400 and 500 pounds with a beautiful black coat. Outside of a grizzly bear in the Yukon, this was the largest bear I had ever seen.

Greg recommended that I contact Phil Smith, the owner of the Bluff Creek Resort at the base of Fish Lake Road. He stated that Phil has some interesting Bigfoot stories that I would appreciate. He also promised to keep his ears open and relay any other information to me when he heard it. Greg took me back to my vehicle. I sincerely thanked him for the trip and his hospitality. I really appreciated my two and a half hours with Deputy Greg Berry, a true professional!

After making my initial contacts with government employees and then introducing myself to a few of the locals, I started to develop some relationships that led me to understanding the habitat of the Bigfoot. It was with the eventual help of locals like Inker McCovey and Ed Masten that I was able to get out into the forests, see the food sources and understand what made Bigfoot keep coming back.

Food Sources

Berries

If there is one natural food crop that flourishes in the Hoopa reservation and at the base of the Go Road it is berries. The bushes can be seen from almost any point in the valley while looking at the valley floor. Many of the bushes easily reach 8–9 feet high and 10–15 feet wide. They line some roads and creeks and act as property lines to some residences. Some of

This is a view across Highway 96 just south of Hoopa looking onto one of the many dirt roads that are lined by huge berry bushes.

the local residents do pick the berries, but many stay on the vine until a bear or Bigfoot gets a grip on them.

I have never been in any region of the country that has such a proliferation of wild berry bushes. One summer while I was working in Hoopa I was walking along a dirt road adjacent to the Trinity River. I was noticing a large amount of fresh bear scat among the berry bushes I was standing adjacent to. As I walked around a corner I came within 25 feet of a 300-pound black bear that was calmly eating berries. He paid very little attention to me and I quickly took some back steps and left the area. The berries are an obvious source of nutrition for the bears. Once you complete reading the sightings section of the book you will understand that the berries are also a large part of the Bigfoot diet.

Matsutake Mushrooms

During the time period from late November until as late as January 31, matsutake mushrooms grow throughout the lower elevations of Hoopa. Matsutake are a delicacy in Japan and their buyers will travel to Hoopa during harvest time and pay as much as $50 per pound for fully developed mushrooms. Many of the locals make several thousand dollars every year picking the mushrooms and selling to buyers.

During my time in Hoopa I had heard many stories about how to find, pick and harvest wild mushrooms. Many of the stories sounded a bit far-fetched and it sometimes sounded a little too easy. On a clear and cold day in early winter I convinced Inker McCovey and Ed Masten to take me to the upper reaches of Mill Creek (outside the reservation) and explain how they find their mushrooms and take me through the steps in harvesting. Since there are several occasions noted in the sightings section where witnesses have seen Bigfoot while hunting for mushrooms, information on how it is done may be helpful in understanding the sighting.

Inker and Ed first explained that it takes moisture, lots of leaves and filtered sunlight to get quality mushrooms. They told me that after the first heavy rains of the season the mushroom will start to form. There needs to be a period of sunlight after the rains to allow the mushrooms to fully develop. During this sunlight period, the mushrooms will grow rapidly. Matsutake tend to grow in bunches. If you find one, be slow and methodic in your search because there are probably others in the immediate area. You need lots of leaves to protect the mushroom in its infancy and also to protect it from snow that will hit the area during these months. While we were on our mushroom hunt there was snow in some of the areas we searched and the leaves did act as a buffer between the mushroom and the elements.

Searching for matsutake is definitely a skill that takes time to develop. After following Inker and Ed for 30 minutes I started to get a feeling for the type of filtered sunlight the mushrooms need. It was also easier to understand the elevation requirement and the type of trees in the immediate area that the mushrooms need — lots of oak trees — and this is why the Natives call the mushrooms "tan oaks." The process of finding the mushroom can take a long time, moving a lot of dirt very carefully. If you move the dirt too fast, you break the mushroom and decrease its value.

A matsutake that is protruding from the ground and is partially covered by snow and dirt can be seen in the middle of the photo.

Once an area of matsutake mushrooms is found, the snow is very slowly moved aside and it usually has leaves frozen to it. As you are moving these aside, small tan spots anywhere from one to three inches in diameter appear and you will notice the mushroom shape. Most people use a small fork-shaped hoe and then wedge the mushroom out of the ground. In an area that is 15 feet in diameter you may be able to find as many as 10 mushrooms. Inker explained that the mushrooms sometimes run in veins across the hillside. Once you find one patch and collect everything around the original area, don't give up. You want to keep collecting in angles out of that area until you lose any sign of matsutake.

Ed explained that his tribe loves to eat the mushrooms. He broke off a small piece and told me to smell it. The mushroom has a distinct smell of oak and it's one you will not forget. He said that the most common method for cooking the mushroom is to slice it thinly and fry it in a pan with butter; it's not healthy, but it's tasty.

Finding areas to collect the mushroom is not easy. It takes a special permit from the United States Forest Service if you want to collect mushrooms in their forests. Only tribal members are allowed to collect mushrooms on Hoopa reservation property and if a non-tribal member is found collecting on tribal property, they will be arrested and prosecuted.

Ed, Inker and I were at approximately 2,000 feet in elevation when we found our matsutakes. While they were patiently and quietly working the soil for the mushrooms I stood back and watched for much of the time. I wanted to understand the circumstances of how Bigfoot would be seen while others were collecting mushrooms.

I should remind you that there are several historical stories indicating that Bigfoot in the Hoopa area regularly eat matsutake mushrooms. Bears and deer also regularly eat the delicacy, and both are also found in abundance in the area. While we were hunting the mushrooms, Inker showed me where a deer had scratched the soil, found a mushroom and took one bite out of it. This was one way that Inker found the area where the mushrooms were growing.

While hunting the mushrooms you are almost always looking down and you are usually alone and not talking. If you are in an area where there should be matsutake mushrooms, you are being quiet, staying low to the ground and engaged in your work, there is a very decent chance that something could walk up on you without it noticing you, or you noticing them. There are many instances where Bigfoot has walked up to people while they were engaged in collecting mushroom. A pile of mushrooms would definitely emit a significant odor and that might possibly lure the Bigfoot into the area. In Inker's statement about his sightings he explained that he had collected two large buckets of matsutake mushrooms, saw a Bigfoot and its baby, and emptied the matsutakes on the ground as a goodwill gesture and left the area. Charles Masten was on a walk two weeks earlier in the same climate and altitude range we were when he saw Bigfoot. If Bigfoot was looking for a food source in the snow, he would be in

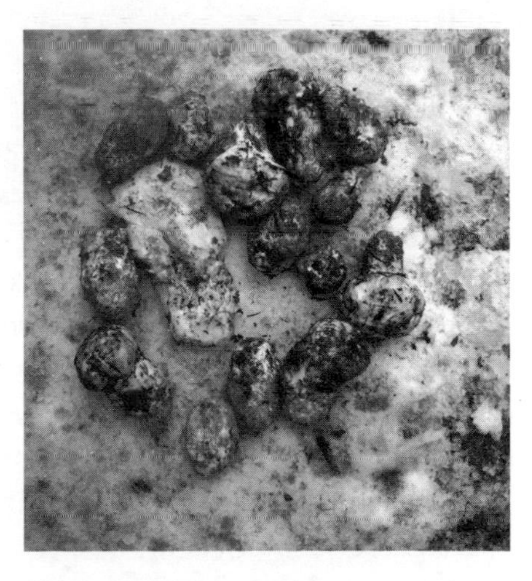

Fourteen matsutake mushrooms collected in one 10-foot diameter location.

the area where we found the mushrooms. It should be noted that the leaves in this area were 4–6 inches thick. When I wiped all of the leaves away I found rock-hard dirt and gravel. It would not have mattered if it was leaves or dirt, Bigfoot tracks in this area would be nearly impossible to find because the ground and its cover are like cement.

Acorns

A product of the oak trees is the acorn. It seems that almost anything living in and on the ground and flying through the air likes acorns. Squirrels, bears, Bigfoot and birds are all in competition for the native nut. In the 1800s and early 1900s Native Americans accumulated the acorns, ground them into a fine powder and made them into an eatable part of their diet. If you pick up and eat an acorn it will taste bitter. It takes some preparation and lifestyle acclimation to appreciate the nutritional attributes of the acorn. The Hoopa reservation has thousands of oak trees, and a small search for the acorns reveals many of them can be found on the ground.

Fish

At the center of the Hoopa reservation is the Trinity River. The river delivers one of the main food sources for the tribe and also one of the primary tourist attractions to the area — salmon and steelhead fishing. The coho and chinook salmon and the steelhead are not the only fish in the Trinity River, but they are considered at the top of the food chain for sport fishermen. The river also contains Lamprey eels, sturgeon, trout and suckers. I will concentrate on the most important fish in the river for the tribe members.

The chinook salmon run the Trinity from April to September and are the biggest salmon on the Pacific coast. The coho run the river from September to December, and are much smaller than the chinook.

The Pacific Lamprey is also a staple to the tribal members. They live

4–5 years in fresh water and then migrate to the ocean, where they stay 6–18 months before returning to the river to spawn.

The green sturgeon is the largest fish in the river. They migrate up the Trinity from February to July to spawn. They get as far upstream as Grays Falls before their trip ends and they spawn out. In later years the sturgeon move to the lower Klamath River estuary before they go out to the ocean.

The biggest obstacle in recent times to the successful spawning and breeding of salmon and steelhead has been low river flows on the Klamath. The salmon and steelhead need cold fresh water to get their eggs hatched and a low river flow means warmer water with less area to lay their eggs. The main reason for low flows on the Klamath has been farmers upstream demanding more water for their crops. The U.S. Department of Reclamation has been in many lawsuits in recent years between the various tribes and the farmers, as each group has been demanding more water.

Relationship of Bears & Deer to Sasquatch

Hoopa Valley Tribal Council
Natural Resources Department

J. Mark Higley
Wildlife Biologist

During my months in Hoopa I had many employees from the Natural Resources group come to me and explain that they wanted to work cooperatively on my project, and others even told me their Bigfoot stories. A few of the employees said that the management of the department had told them to steer clear of me and not dispense any information or photographs. The more time I spent in Hoopa the more my name became circulated, and fortunately for me, I started to garner more credibility and friendships.

To understand Bigfoot it is important to understand its food source (deer) and competitors for its food (bears). If you have an understanding of how these animals migrate, find their food, where their food is located and the amount of food available, then you will have a better understanding of the creature itself. I was extremely fortunate to study an area such as Hoopa. Considering the Hoopa tribe owns 144 square miles, it has a relatively large staff that manages their wildlife and forests. In contrast, the local United States Forest Service (USFS) Office in Orleans has a smaller staff that is responsible for an area twenty times the size of Hoopa, and the USFS staff doesn't have the same vested interest in preservation (dollars into their bank account) that the Hoopa staff possess. The Hoopa tribe directly profits from the sale of their forest products; the better they manage, the more they make as a tribe.

During one of my visits to Hoopa Natural Resources I walked into the office of the group manager, Mark Higley. I explained to Mark that I'd like to talk to him about bears and dccr, and asked if he could offer me some time. Mark pulled up another chair and said he had fifteen minutes.

I had heard from locals that there was a specific deer migration that started from the southern part of the reservation, went north and then crossed the Trinity near Mill Creek and onward towards the coast. I asked Mark if there was a migration that followed the path described. Mark replied that there was a migration, but not as specific as I explained. He stated that in the middle of winter the bucks (male deer) stay just below the snow line at approximately 4,500 feet. The deer on Mill Creek Ridge, Tish Tang Ridge and Big Hill Ridge all move from the high country in the summer to lower elevations in the winter. He stated that the move is subtle and they usually do not cross rivers going from the eastern ridges to the coast. On the western hills of the reservation the migration of deer is almost non-existent.

I asked Mark to give me his opinion about the numbers of deer on the reservation. He explained that there are few deer on the reservation compared to property adjacent to it. Mark attributes the low number to year-round hunting, few regulations for tribal members on hunting and high hunting pressure. Tribal members can hunt from their vehicles 24 hours per day, 365 days per year and can shoot male or female deer, all contrary to California State law outside the reservation. Mark explained how the wildlife group attempts to monitor deer numbers by utilizing game cameras mounted on trees. The Natural Resources Group learned years ago that they would not get many photos if the cameras were within two hundred yards of the dirt roads because that was the target range for tribal members who were hunting. They had to mount the cameras far inland from the roads to catch any deer on film, which they did; some huge doe and bucks were caught on film.

One of the most controversial subjects in Bigfoot research is the alleged correlation between bears and Bigfoot. There have been many Bigfoot researchers who have stated that Bigfoot is located where bears are located — the more bears, the more Bigfoot. There may be some rationale to the theory in that the more bears, the more food source, the more food source then more Bigfoot. It does appear that Bigfoot and bears share the same dietary practices in Northern California.

According to the United States Forest Service, there are approximately 600,000 black bears *(ursus americanus)* in the United States. These bears range from 100 to 200 pounds for females and 350 to 700 pounds for males, with the average male in the 400-pound range. The black bear can live to 30 years old if not under heavy hunting pressure. The Forest Service states that under their best estimates, there are 20,000–24,000 black bears in California. The bears have a history of eating grasses and berries, and feed off the kills of other animals. Black bears have also been known to take down small deer, although this is rare.

Mark and I now started to discuss the bear population in Hoopa. Mark immediately stated that after he and I were done talking that I should also speak to one of his assistants and project manager on bears and deer, Elaine Creel. I told him that I appreciated the recommendation and we continued with the conversation. Mark told me that he had been a biologist for Hoopa since May of 1991 and he had seen populations of bears increase while the population of deer decreased.

Mark said that the best bear consensus estimates that they possessed stated that there were 400–600 black bears on the reservation for 144 square miles. He explained that some areas of the reservation had triple the population of other areas, with the eastside having the fewest number of bears. He said that 80 percent of the bears were black bears, 15 percent brown or cinnamon, and 5 percent blond (a genetic anomaly).

I asked Mark about the rumors that I had heard that the bears in the valley do not hibernate, and if that was true. Mark stated that the healthiest bears do hibernate and that females that are pregnant must hibernate. He stated that the biggest bear they had ever captured was 400 pounds, but they do get larger than that. The bears in the valley may den only a few months, and if the food source is available year round they may not den at all. He also stated that most bears take short naps.

I now asked Mark to explain the food sources for the bears in Hoopa. Here's what he told me:

> **April:** Acorns left over from earlier season, some may be rotten.
> **July:** Berries, cherries, plums, apples, anything from fruit trees.
> **Year round:** Fish from tribal members' fishing nets (learned behavior from watching the fishermen clean out their nets).

This is a behavior that has been learned over the years. Members have joked that the bears have become so good at this that they steal the fish and even re-set the nets for them.

I also asked Mark if they get many calls from tribal members about issues with bears and how to control problem bears. Mark said this is a constant headache and they are always answering questions about how to manage bears. He stated that there was one example where a tribe member called and said that they had a goat tied to a fruit tree. A bear kept harassing the goat in its attempt to get to the fruit that was falling on the ground. The Wildlife Group explained to the member that they had to move the goat or risk losing it to an attack. The tribe member didn't take the recommendation, and the bear killed and ate the goat. Mark also noted that tribe members do not generally kill bears. Certain tribal members feel they have evolved from the bear and are tied to them in some evolutionary way, thus, there is a huge population of bears and few deer.

Elaine Creel
Project Manager, Hoopa Wildlife

The day after meeting with Mark I met with Elaine. She is a 29-year old graduate student at Humboldt State University and a project manager on a federal grant to study bears, deer and woodpeckers. I stayed clear of the woodpecker questions.

Elaine started the conversation by saying that she spent her earlier years studying on the Channel Islands in Southern California and it was nice to be on land and studying wildlife.

The first question I asked Elaine is why Hoopa has such a large population of bears. She explained that the tribe had specific timber practices that aided the bears in their growing population. The tribe steered clear of harvesting hardwoods as the hardwoods produced the acorns that bears ate. Elaine also stated that tan oak acorns were one of the bears' main food sources and there were significant numbers of these on the reservation.

I asked Elaine to tell me about the tribe's bear management program and how they kept track of bear migration and problem bears. She stated that they had 10 bears wired with GPS collars. They can download bear locations at anytime and are always watching and trying to understand their movements. At this point one of Elaine's co-workers had heard our conversation and brought in a GPS printout of the bears monitored and their associated movements. It was immediately obvious from the printout that many of the bears lived adjacent to the Trinity River, and none had crossed the river during the year. I now asked about the bears and the river and their association. She said that they just finished doing genetic testing on the bears and found that the animals do not appear to cross the river. This surprised the group, as it was not obvious that the Trinity River is a natural divider in bear populations.

I was further told that they find more cinnamon and blond bears in the eastern hills. She added that while the population of these colored bears is relatively low, they could be easily located.

In all my time in Hoopa I had heard only one story where a bear was moderately aggressive and it wasn't towards a human, it was the killing of the goat that Mark had described. I asked if there had been any reports about a black bear being aggressive towards a human. She said no.

The conversation with Ms. Creel centered mostly on bears, but I ended with a few questions about deer. I asked her how she felt the deer population in Hoopa had changed in the last 20 years. She stated that the population of deer had definitely gone down, explaining that she had talked to elders about deer and they had told her that 20 years ago you could find deer on the valley floor, but now you never see them. She did specifically state that the lack of deer is not due to a lack of food source.

Correlating the Data

After looking at the Hoopa tribal website, researching United States Forest Service documents and reading academic articles on bears, there are some very interesting correlations that can be made about bear populations in California compared to bear populations on the Hoopa reservation.

Location	Number of Black Bears	Available Acres	Square Miles	Bears Per Acre/Sq Mi
Hoopa	600 (Max)	89,572	144	1 per 149/4.1 Sq Mi
Hoopa	400 (Min)	89,572	144	1 per 224/2.8 Sq Mi
California	24,000	(Max)	18,500,000	1 per 771
California	20,000	(Min)	18,500,000	1 per 925

Number of black bears:

- Hoopa Tribal Wildlife estimates they have 400–600 black bears on 89,572 acres in their reservation.
- The United States Forest Service estimates there are 20,000–24,000 black bears on available land that could be forested (18,500,000 acres).
- The data indicates that Hoopa has over four times the number of bears as the rest of California using the minimum correlating data numbers, and five times the bears using the maximum numbers.

The California Department of Fish and Game has been compiling bear-kill statistics by county for the last 10 years, 1996–2006. The graph on page 70 indicates the eight counties in the state with the highest numbers of bears killed in the last 10 years. I have also searched the three most popular sources for Bigfoot sightings reports and listed the total number of sightings reported to them for the same time period. It's interesting to note that the United States Forest Service and the California State Fish and Game differ in their estimates of black bears in the state. Fish and Game estimates in

2007 there were 25,000–30,000 bears alive and well in the state. They estimated in 1982 that there were 10,000–15,000 bears in the state; so, the bear population in California has doubled in 25 years. Fish and Game also estimates that bear densities range from 1.0–2.5 per square mile throughout the state where the area has forests and is habitable by bears. Fish and Game also says that more than half of California's 25,000–30,000 bears live in the north coast/Cascade region, and thus the high kill ratios in those counties. The last year that California Fish and Game had data was in 2006, indicating that they issued 21,786 bear tags (authority to shoot and kill a bear) and 1,735 bears were taken, a success rate of 7.9 percent. Reports indicated that it took the successful hunter four days to bag their bear.

County	Bears Killed	1996–2006 Yr Most Bears Killed/Number	Yr Least Bears Killed/Number	Bigfoot Sightings
Trinity	2015	1999/242	2002/179	10
Siskiyou	1844	1999/260	2005/121	13
Shasta	1661	1999/208	2001/104	10
Humboldt	1310	1999/164	2005/98	28
Mendocino	891	2001/122	2005/56	4
Tulare	1077	2003/158	2005/62	8
Fresno	789	2000/94	2005/59	6
Plumas	769	1997/98	2000/49	4

Evaluating the Data

It's a fact that Hoopa has significantly more bears per acre than anywhere in California and significantly fewer deer than other Forest Service property in other parts of the state. It's also a fact that bears in Hoopa are not hunted, they have large quantities of food, and California Fish and Game has no authority on the reservation, thus no depredation permits.

A county-by-county search through California Fish and Game archives indicate that Humboldt County sits in the number five position for most bears killed in a California county from 1996–2006 (see chart). In that same

10-year period, Humboldt County has more than double the amount of Bigfoot sighting reports than any other county in the top eight in bears killed.

The question that science needs to ask is, "Is there is a correlation between the number of bears in a specific area and the number of Bigfoot?" Knowing that Hoopa and its surrounding region has had a large number of Bigfoot sightings, it makes the bear–Bigfoot connection compelling. Does Bigfoot know that it will not be harmed if it stays confined to the area around Hoopa? Does it realize that the Hoopa people will not shoot at it?

There is another area that sits next to the Hoopa reservation that has an equally compelling case for Bigfoot — Redwood National Park. The park sits just west of Hoopa and shares Bald Hills Road as a commonality. Redwood National Park does not allow hunting, has few guests in comparison to Yosemite and Yellowstone, and yet has a huge name in the Bigfoot community as harboring a high concentration of Bigfoot. The park does sit in Humboldt County and is an area that has never had significant foot or vehicular traffic. It should be noted that Bald Hills Road has its start at the entrance of the park and is famous for Bigfoot sightings. (See "Sightings" and "Incidents" chapters for more details.)

Chapter 3
Bigfoot Sightings Overview

I have documented a combined total of 48 Bigfoot sightings and incidents in and around the Hoopa reservation. I have purposely put the list into two categories for a variety of reasons. "Sightings" are instances where the witness has claimed specifically to have seen a Bigfoot. An "incident" is where the witness has claimed to have observed activity related to that of a Bigfoot. Sighting reports require the witness to sign an affidavit; incidents generally do not. However, some of the incidents were very unusual, and in those cases I did request that an affidavit be signed. Each list was categorized starting with the oldest sighting or incident and going forward to the most recent.

This list represents an exhaustive effort to find credible witnesses and confirmed Bigfoot sightings. The list has been studied for many months and numerous interesting facts have emerged. It should be emphasized that almost all third-hand stories have been excluded. The Georgia Campbell story is included because of the odd location and because it was her niece who told us of the incident.

The first oddity of the list is the break in sightings from 1988–1996. It was almost as though Hoopa had folded up and disappeared from Bigfoot visits for 12 years. There is also a significant nine-year time break from 1975–1984. The monthly breakdown of sightings doesn't indicate any two consecutive months where there are no Bigfoot sighting reports.

Monthly	Bigfoot	Sightings	List
January	0	July	4
February	2	August	4
March	0	September	2
April	1	October	1
May	3	November	7
June	0	December	1

The one significantly high month for Hoopa Bigfoot sightings is November. The only important food source that flourishes in that month is a high number of fish in the river. There are large quantities of salmon and steelhead that are simultaneously occupying the Trinity and taking

Sightings List

No.	Date	Witness Name	Time	Location	Description
1.	1956	Georgia Campbell	Noon	Bull Ranch	Mom and smaller Bigfoot
2.	5/63	Josephine Peters	Noon	Slate Creek	Bigfoot in creek
3.	1964	Inker McCovey	8 pm	Residence/Mill Crk	Arm reaches into room
4.	1967	Jesse/Nancy Allen	10 pm	Forks of Salmon	One Bigfoot/field
5.	7/72	Michelle/ Leanne McCardie	1pm	Shoemaker Rd	Adolescent Bigfoot
6.	7/75	Jackie Martins & Friend	9 pm	Bald Hills Rd	Bigfoot crosses road
7.	8/84	Tane Pai-Wik	8 am	Pecwan	Bigfoot looks in window
8.	1984	Alice Barker	9 pm	Bald Hills Rd	Bigfoot crosses road
9.	11/84	Damon Colegrove	6:30 am	Somes Bar/Camp 3 Rd	Bigfoot seen
10.	7/85	Lillian Bennett	10 am	Bald Hills Rd	Bigfoot crosses road
11.	1987	Phil Smith	Noon	Bluff Crk Resort	Bigfoot seen
12.	10/88	Jay Jones	12:01 am	Red Crk Valley	Bigfoot seen
13.	8/92	Joe O'Rourke	10 pm	Pecwan-Banks of Klamath	
14.	8/96	Juliene McCovey	6:30 am	Tish Tang Valley	Female Bigfoot seen
15.	9/03	Ed Masten & Friend	4 pm	Upper Tish Tang	Bigfoot seen
16.	2003	Page Matilton	Noon	E/O Willow Crk on Trinity River	Seen
17.	11/03	Inker McCovey-Mary McClelland	12:01 am	96@Mill Crk	Seen running
18.	12/03	Clifford Marshall	9:30 pm	96 N/O Hoopa	Seen Jumping/cliff
19.	2/04	Mary McClelland/ Sadie McCovey	1 pm	Berry Summit	Chasing Deer
20.	4/04	Debbie Carpenter	6:30 am	Carpenter Lane	Seen w/berries
21.	8/04	Raven Ullibarri	Midnight	Upper Mill Crk Rd	Bigfoot in garbage
22.	9/04	Raymond Ferris	Midnight	Upper Mill Crk Rd	Squatting in Rd
23.	11/04	Michael Mularkey	4 am	96at Shoemaker Rd	Running across Rd
24.	11/04	Inker McCovey	2 pm	Mill Crk at Big Mtn	2 Bigfoot seen
25.	5/05	Gordon McCovey	5:30am	Dillon Crk at Klamath	Bigfoot walking
26.	5/05	Paul James & Richard Nixon	8:30 pm	169 N/O Weitchpec	Crossing Rd
27.	7/05	Pliny McCovey	8 pm	Upper Mill Crk Rd	Big & small Bigfoot
28.	8/05	Kim Peters	11:30 pm	Mill Crk Rd shortcut at 96	Seen
29.	11/05	Ed masten	Noon	Upper Tish Tang Rd	Bigfoot Seen
30.	11/05	Romeo McCovey	8 pm	96 at Mill Crk Rd	Seen near creek
31.	11/05	Jay Jones	10 pm	Upper Tish Tang Valley	Bigfoot Seen
32.	2/06	Josephine Peters	9 pm	Residence/Loop Rd at Supply Crk	Seen
33.	11/06	Charles McCovey	1 pm	Upper Mill Crk Rd	Seen walking

Total Sightings = 33

adjacent feeder streams (Mill Creek, Supply Creek, Tish Tang, Camel Creek). Since November is the first month when the river has a high number of these large fish, this may account for the large number of Bigfoot sightings. The location of the fish in the feeder streams may also be a reason why you will later see that there are large numbers of sightings near these listed creeks. Perhaps Bigfoot fills up on fish in November and then gets tired of that diet. November is also one of the first months that bears would hibernate in the area, if they do.

Times of Bigfoot Sightings Chart

Midnight	3	5am	1	10am	1	3pm	0	8pm	4
1am	0	6am	3	11am	0	4pm	1	9pm	3
2am	0	7am	0	Noon	5	5pm	0	10pm	3
3am	0	8am	1	1pm	3	6pm	0	11pm	1
4am	1	9am	1	2pm	1	7pm	0	Midnight	3

The time graph of a Bigfoot sighting shows that few are seen between 1:00 a.m. and 4:00 a.m. The logical answer to this is that there are very few people from Hoopa out during that time of night. The one sighting at 4:00 a.m. was a supermarket manager on his way to work. Many Bigfoot researchers believe that the creature is nocturnal. I am not sure; however, our data in Hoopa indicates that the creature has a preference of night versus day. The category of 7:00 a.m.–7:00 p.m. includes 13 sightings of the 32 total. During the summer the days are longer and we can logically extend daylight hours till at least 8:00 p.m., and that would raise the daylight sighting numbers to 17, more than half of the 32 total sightings. But we do have to reconcile the fact that a vast majority of people are not outside from 1:00 a.m. to 5:00 a.m. and thus a Bigfoot sighting could not be made. From noon to 3:00 p.m. there are nine sightings reported. The answer to this may be that people are on their lunch hour and out in the woods, but I doubt that accounts for the entire reason. This region gets significant rainfall during winter months and as such the sky is cloudy a lot during the day. If Bigfoot tries to avoid bright light and the heat of the day, this may be the reason it enjoys Hoopa and the north coast. I think it is interesting that during the winter the sky does get dark early at 5:00 p.m. The timeframe from 5:00 p.m. to 7:00 p.m. would be a time when a lot of people are on the road going home from work, going to the store, etc. There are no sightings reported during that

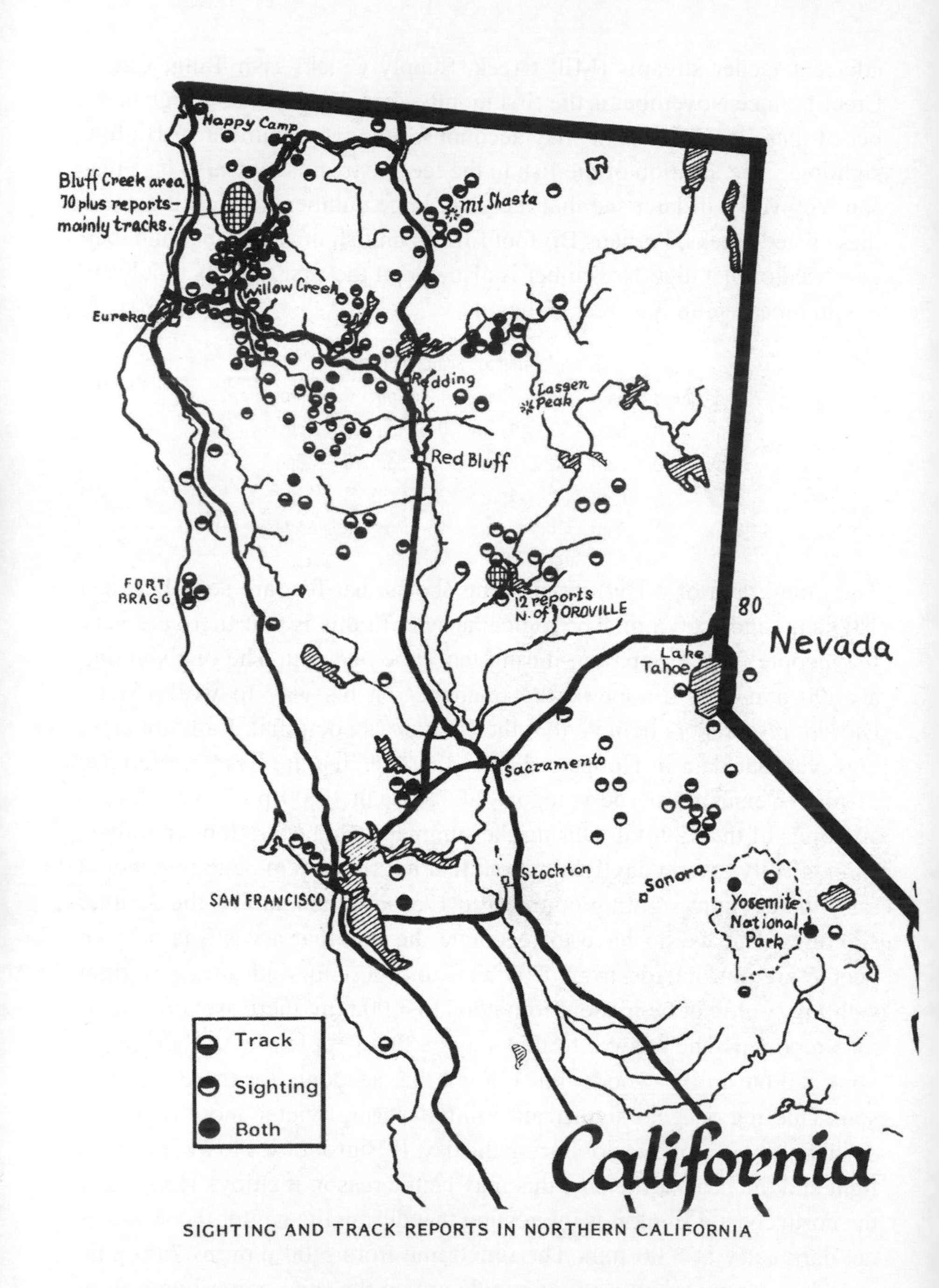

SIGHTING AND TRACK REPORTS IN NORTHERN CALIFORNIA

time frame. It might be possible that Bigfoot has learned to stay away from the roads during the dinnertime commute.

Bigfoot Incidents

A witness to activity or behavior that is likely attributed to Bigfoot is included in this category. The activity cited in this category may be rock throwing, screaming, odors, footprints, bipedal stalking on a trail and any other physical evidence likely left by Bigfoot.

No.	Date	Name	Time	Location	Activity
1	8/62	Hank Masten	10 pm	Mill Creek Lake	Screaming
2	1964	Corky Van Pelt	9 pm	Camel Crk at Trinity River	Rocks
3	7/69	Carlo Miguelino	2 pm	12 miles e/o Big Hill Rd	Screams
4	8/71	Hank Masten	Midnight	Trinity Rvr at Tish Tang	Rocks
5	1975	Jeff Lindsay	9 am	Papoose Lake	Footprints
6	7/80	Jeff Lindsay	Noon	Enni Camp	Footprints
7	8/80	Dave Bishop	2 am	Red Cap Creek	Footsteps/scat
8	1994	Corky Van Pelt	4 pm	Upper Camel Crk	Footsteps-Noise
9	2002	Phil Smith	3 am	Bluff Creek Resort	Screams
10	4/04	"Winkle" White	2 pm	Trinity at Carpenter Ln	Footprints
11	1/05	Hank Masten	5 am	Lower Mill Crk Rd	Footsteps
12	5/05	Doreen Marshall	9 pm	96 at Mill Crk Rd	Something seen
13	2/06	Carlo Miguelino/son		2 pm	Big Hill Rd at reservation border Print
14	6/06	John Harlan	3 am	Aikens Creek Campground	???
15	8/06	Warrior Sanchez	9 pm	Transmitter Hill	Rock Throwing
16		Two incidents are documented under Warrior's Statement			
17	10/06	James Marshall/ Christy Brown	3 am	Transmitter Hill	Footsteps
Total Incidents 17 Total Witnesses 20					

There has been a large increase in Bigfoot incidents in the last years. Incidents were fairly consistent at two per decade for the previous 40 years and then a steep escalation occurred. The most common incident is seeing

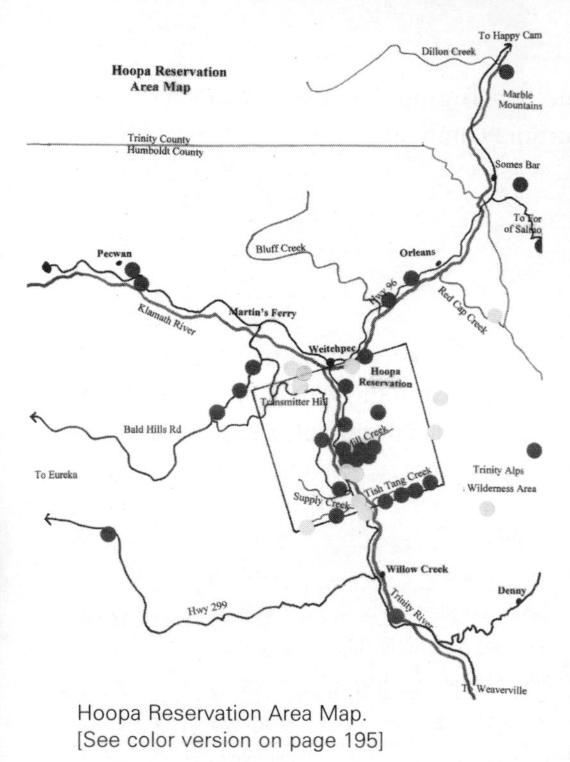

Hoopa Reservation Area Map.
[See color version on page 195]

footprints or hearing bipedal steps. Seeing the footprints is significant in this region because almost everyone in the reservation knows the difference between a Bigfoot print and a bear print. A visitor coming from the city may mistake a large bear print for a Bigfoot print, but I do not think any seasoned tribal member would make that mistake.

There are several incidents where Bigfoot has thrown large boulders in the general direction of the witness, all being thrown from a ridgeline. None of the boulders get close to hitting the people, but they are close enough to catch their attention. The boulders that are thrown are huge, some over 100 pounds, all being thrown in the general direction of the witness. I doubt there is anyone in the world that could throw a 10-pound rock 200 yards, yet people are claiming that they see 100-pound boulders flying over 200 yards in the air. The only scenario I could imagine (prior to investigating Bigfoot) where a boulder of 100 pounds could be thrown 200 feet would be a catapult or volcano, not by any animal or human. Many of the rock-throwing incidents are in conjunction with screaming that witnesses state does not come from any "normal" animal in the forest. The Camel Creek/Tish Tang and Trinity River region was a very hot spot for Bigfoot incidents.

Activity for the Last 40 Years

The Hoopa reservation sightings map [see color version on page 196] is an overview of the entire region where I investigated Bigfoot sightings and incidents. Yellow dots indicate the location of Bigfoot incidents and red are actual sightings. I decided to use the entire area around the reservation to collect sightings because Bigfoot is a very mobile creature that can migrate and wander tens of miles per day. There does appear to be a large accumulation of sightings and incidents inside the reservation. A skeptic may claim that the reservation is where the population is, thus the sightings would likely be made there. The irony of the map is that Willow Creek claims it is the Bigfoot capital of the world, yet I couldn't find one witness to a sighting or incident inside of the Willow Creek limit or vicinity. The closest witness I could find was a Hoopa teacher who saw a Bigfoot in the Trinity River outside of Willow Creek. Bluff Creek is located at the top of the map and is obviously famous in Bigfoot lore, yet I couldn't find anyone who had seen or heard of any recent activity in that area. The sighting at the far top right corner of the map at Dillon Creek was made by a Hoopa tribal member and is within the normal range of their travels. The sightings on Bald Hills Road are just on the outside fringe of the reservation.

The Hoopa reservation map highlights the sightings made inside the reservation. Most of the sightings are in the north end near Mill Creek Road. The sightings in this area are not a new occurrence; they have been happening since the 1960s. This map makes it appear that many of the sightings are located adjacent to a creek or valley, which is reality. This may be occurring because that is where the roads and residences are located and people are visiting and lounging. The other viable reason for the creek or valley sightings is that this is where Bigfoot is finding food sources — fish, berries, grasses, etc. I can attest that the creek area above the residences on Lower Mill Creek Road have huge quantities of berries growing along its bank. I have also been in this area while the salmon

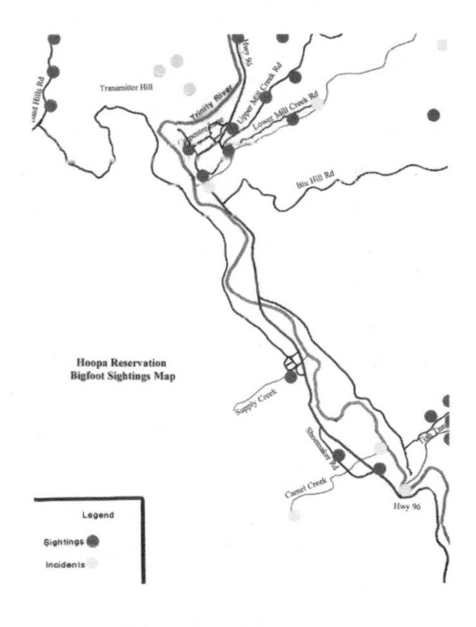

Hoopa Reservation Sightings Map. [See color version on page 196]

and steelhead are in the rivers and there are large numbers of fish that are breeding in upper Mill Creek; it's a beautiful sight.

Bigfoot Alley

This is an aerial view of the region on the north end of the urban area of the reservation. The Trinity River is in the middle of the picture, Mill Creek is just inside the lower black line and Transmitter Hill, and the region inside of Bald Hills Road is at the top left of the photo. I have coined this region "Bigfoot Alley" for an obvious reason. Transmitter Hill has a ridge that slowly drops in elevation until it meets the Trinity River in the area where Winkle White had a screaming incident and saw a footprint. The area of Mill Creek Road has a ridge above it that also drops until it disappears at Highway 96 and Carpenter Lane. If the statement of the Yurok "Bigfoot is a ridgewalker" is correct, then this area is prime for an interaction. Charles and Inker McCovey each had sightings at the far right side of this region, and both were either on or very near a ridge. The reasons why Bigfoot frequents this specific area are easy to understand. There are huge berry bushes that line the streets of Carpenter Lane, Mill Creek and the Mill Creek neighborhood. The Trinity River and Mill Creek each have runs of salmon and steelhead. Local fishermen have nets set specifically at the location where Winkle White saw the footprint and heard the screams. It would seem appropriate that if a Bigfoot were to cross the Trinity River, this would be an optimum location, ridgeline to ridgeline.

The area to the left side of the river is the area you would travel west and end up in Redwood National Park. If a creature wwas fleeing horrible conditions in the coastal region this would be an avenue of escape. In reverse, if Bigfoot was avoiding adverse conditions in the Trinity Alps of the Marble Mountains, this would be the area where it would have the first opportunity to cross the river without being in a heavily congested urban environment (downtown Hoopa). If the creature was attempting to cross the river in the northern portion of the reservation or further north of the top of the aerial view it would have an extremely difficult proposition. The area north of the picture is ladened with cliffs plummeting from Highway 96 down to the river. The sighting report from witness Clifford Marshall indicates that Bigfoot will leap off the cliff if forced, though that probably isn't its normal behavior. The area I have described by Carpenter Lane would again be the optimum place to cross if you were traveling south from Weitchpec and were looking for an easy access point to each of the hillsides (Mill Creek and Transmitter Hill).

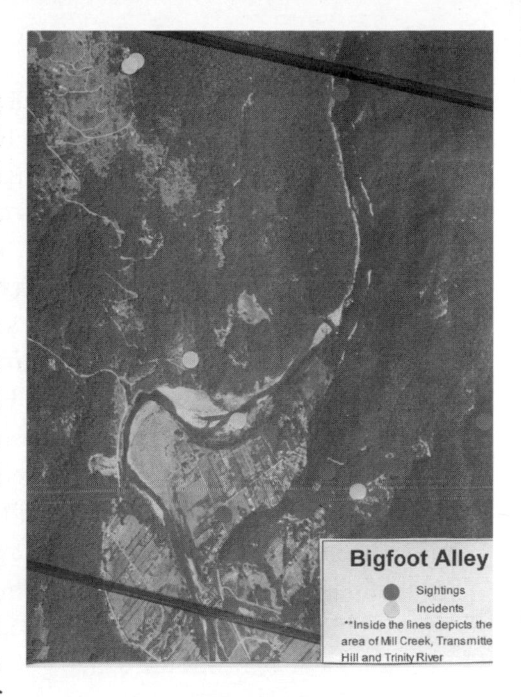

Aerial Photo of Bigfoot Alley.
[See color version on page 197]

Bigfoot Alley has 14 of the 33 Bigfoot sightings documented, which gives it 42 percent of all sightings. It has five of 15 of the Bigfoot incidents, or 33 percent of all incidents. These are astronomically high numbers considering the relatively small area it encompasses on the reservation. I seriously wonder if there is any another place in the world like Bigfoot Alley. I believe that this region deserves a serious scientific study with a team committed to staying for the long haul over several seasons, and with a large contingent of fourth generation night optics to watch the crossings and document the behavior. It would be fascinating.

The region just to the east of the reservation boundary and also east of the outside perimeter of Bigfoot Alley is the Trinity Alps Wilderness Area. This region has always had a high incidence of Bigfoot sightings based on the numbers of visitors who walk in, which is few. There is one Bigfoot incident report on the BFRO.net website regarding the Trinity Alps that made me take notice.

The report states that there were approximately 10 people that took pack mules in the Trinity Alps Wilderness Area in September of 1980. They were there for a 10-day deer hunt of the region with some camping and outdoor enjoyment. It was a four-hour trek into the wilderness before they set up their camp.

After the first day of not seeing any bucks, two individuals decided to set up a deer blind on a nearby ridgeline that occupied a small saddle. After setting the blind and getting into their bags for the night, they started to hear a series of loud screams coming from an area of trees 100 feet below them. The witnesses described the screams as something similar to a sound of a woman who was being murdered. The screams continued for several sequences until they ended with a very deep and different type of scream.

Once the screaming stopped for several minutes they then heard their mule emit a horrendous noise. The animal was located approximately 100 yards from them on a small meadow. Each camper grabbed a rifle and went to the mule's aid. They had small flashlights and didn't see anything obviously wrong other than the mule's knee was swollen. They walked the mule to a location near their tent and all of them then had a peaceful night.

The next morning the mule was examined and they found that its knee was almost double its normal size. This is quite unusual, as it is physically impossible for a mule to twist and injure its knee on its own. Mules are very stable creatures. It was the opinion of the others that something with hands tried to break the mule's leg and in the process of twisting, injured its knee. After a few days the mule was better and could continue its hauling.

The report has several aspects to it that fall into the categories and hypothesis I have outlined in this chapter. The two individuals that left their base camp and set up a satellite camp on a ridge had obviously disrupted the lifestyle of a Bigfoot, thus the screams. The area they were hunting is east of the reservation and is an area near where they recently had fires. The report is fairly recent, 1980, and the geography of the area hasn't changed substantially. The encounter for the campers also falls into another belief I hold about Bigfoot. It is reluctant to approach or make itself known to groups of three or more. In this incident the Bigfoot was making itself known; people weren't leaving, so it appears that it took a little vengeance out on their mule, thus its hysterical state. It was almost as though the Bigfoot knew that when people get into their tents at night, they aren't leaving. These people were getting into their tent when the attack on the mule took place.

I believe that most hikers do not camp on a ridge. Ridges generally have more wind, more inclement weather and usually do not have shelter or water for campers. I believe that if researchers camped on the ridges and looked for wildlife paths and stayed in groups no larger than two, more Bigfoot encounters would occur and more information would be developed. I find the witnesses' story quite believable.

Bigfoot activity really started to accelerate in Hoopa beginning in September 2003. I conducted an exhaustive periodical search of the Hoopa area for this timeframe and came across the following article.

North Coast Journal Weekly ran the following story on September 25, 2003:

> FIRE AND MORE FIRE. Great swaths of Southern Humboldt were shrouded in smoke last week, as two wildfires destroyed over 11,000 acres. Though it was the smaller of the two, the Canoe Fire in Humboldt Redwoods State Park, proved more troublesome. Just as the *Journal* was going to press, the California Department of Forestry ordered a "precautionary" evacuation of the Salmon Creek area. The larger Honeydew Creek Fire broke out in the Kings Range and for a time looked to threaten Shelter Cove; CDF spokesperson Ernie Rohl said that the fire was "behaving itself" Tuesday. This week's heat wave didn't help matters, though. Meanwhile, a Trinity County fire that took out 3,700 acres and closed Highway 299 several times last week was contained The fires have also created "progressively deteriorating air quality conditions" throughout southern Humboldt and western Trinity counties, warned the North Coast Unified Air Quality Management District.

The article is an interesting look at what might have caused the increase in Bigfoot activity. Southern Humboldt County has always had Bigfoot sightings and incidents. Maybe Bigfoot was trying to escape the smoke and fire by heading north into Hoopa. Humboldt Redwoods State Park has a long history of Bigfoot sightings. If Bigfoot is like every other animal in the world, it will flee a fire area. If there was a fire in the park, Bigfoot would flee the area. There is only one direction to run from this park, as it's on the ocean, and you must run east towards Hoopa. In this scenario you have animals leaving southern Humboldt County for areas north. You have animals leaving western Humboldt County and going east, all points towards Hoopa.

I have interviewed many local elders and tribal members about their belief in the upswing of Bigfoot activity. All of the people I have interviewed about this specific reason point to one reason and one fire — the Megram Fire in 1999.

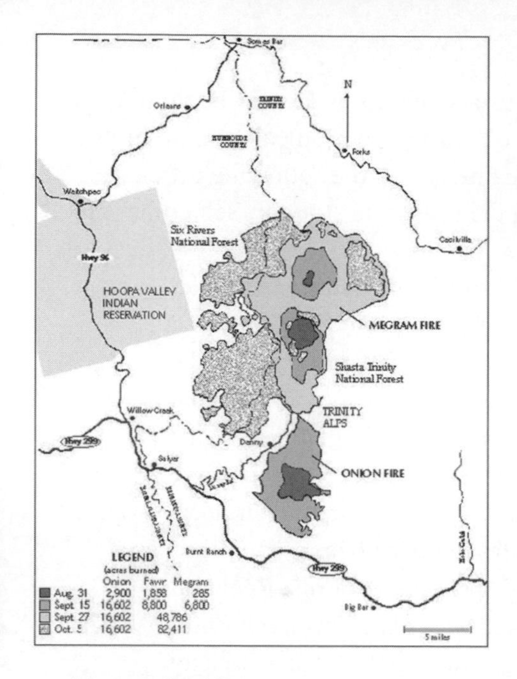

Megram Fire Map.

I think the issues that are causing the escalation of Bigfoot sightings and incidents are multiple events that put Hoopa in the center of the mix.

The Megram Fire was one of the worst that ever occurred in the Trinity/Humboldt County area. The fire burned uncontrolled for many days and eventually destroyed 125,000 acres. As a comparison, the Hoopa reservation has 88,000 acres and is 12 miles by 12 miles square. Many of the older members of the Hoopa tribe feel that Bigfoot lived in the outskirts of the Trinity Alps Wilderness Area and walked into Hoopa at various times of the year for food. They told me that parts of the wilderness area are so remote and difficult to travel that nobody expends the effort to see it. They state that this isolation makes a great home for Bigfoot. Inker McCovey is emphatic that the Megram Fire is one major cause for Bigfoot staying closer to Hoopa. Inker said that the fire destroyed so much acreage that Bigfoot will not be able to return for 10–15 years. The Megram Fire, combined with several other fires in the immediate area over the past 7–9 years, left behind significant destruction.

If you take into consideration the fires that have enveloped the region near the Megram Fire (refer to Megram Fire Map) and couple that with the fires and smoke that occurred in September 2003, you have several dangerous conditions that leave Hoopa as the one safe haven. Hoopa has a large fire response team that is permanently stationed in the middle of the reservation. It is huge when you compare it to other United States Forest Service fire response teams in adjoining cities and counties. Hoopa has the team to protect the city, its citizens and the infrastructure because the livelihood of the tribe is based upon its ability to work and harvest the land. Maybe Bigfoot has been taught a behavior that it will always be safe from fire if it stays in the Hoopa Valley region and reservation. Bigfoot obviously doesn't see the reservation boundary painted on the ground, but maybe it has learned that this area of Humboldt County doesn't burn like other regions.

It is important to remember that the Hoopa valley isn't a common topographical occurrence in this region of the United States. Valleys do not occur much in this mountainous region of Northern California and as such may be an interesting anomaly to Bigfoot. The area in Hoopa, around Hoopa and in the wilderness areas reveals fascinating behavioral aspects of Bigfoot. The stories of the witnesses are mesmerizing.

No.	Sighting	Location	Elevation
1	Patterson Gimlin Bigfoot Film	Bluff Creek	2300 Feet
2	Multiple Sightings, BFRO Website	Fish Lake Campground (Between Bluff Creek and Klamath River)	2300 Feet
3	Bigfoot Nest	Blue Creek	2300 Feet
4	Bigfoot Nest	Bald Hills/Transmitter Tower Rd	2300 Feet
5	Ed Masten Sighting	Upper Tish Tang	2300 Feet
6	Inker McCovey Sighting	Upper Mill Creek	2300 Feet
7	Alpine "Zoobies" BFRO Website	San Diego City	2300 Feet
8	Mary Mclelland/Sadie McCovey	Near Berry Summit	2300 Feet
9	Prior to Summit–BFRO Website	Buckhorn Mtn 1970s	2300 Feet
10	Jerry Crew	Red Cap Creek Bed	2100 Feet
12	Dave Bishop	Red Cap Creek Incident	2000 Feet
13	Charles McCovey	Upper Mill Creek Plateau	2200 Feet
14	Damon Colegrove sighting	Somes Bar	2100 Feet
15	Carlo Miguelino Incident	Big Hill Rd	2400 Feet
16	Jackie Martins sighting	Bald Hills Road	2400 Feet
17	Lillian Bennett Sighting	Bald Hills Road	2200 Feet
18	Juliene McCovey	Upper Tish Tang sighting	2200 Feet

Elevation Association

When I first started to investigate and conduct research into Bigfoot, I started from the onset to look for patterns or similarities in the sighting groups. After my first trip to Hoopa I came home and purchased a wristwatch that indicated elevation. Every time I went on an excursion to visit a sighting or incident location I took the elevation. Almost from the beginning I saw that there were associations based on elevation.

In California, snow usually starts to fall at the 2,000-foot elevation but it usually doesn't stay on the ground very long. The 2,000-foot point is consistent whether you are referring to the coastal mountain range or the Sierra Nevada Mountains. One of the first locations I ever checked for elevation was the original site where the Patterson–Gimlin Bigfoot film was shot. I had reviewed topographic maps of the area and found that it was 2,300 feet and just inside Del Norte County.

I went on to review a vast number of sightings throughout California and found that many were in the range in elevation of what I have listed above. Most of the sightings on all other databases do not list elevations, but in some sightings you can determine elevations based on concise directions. I would encourage anyone with even a casual interest in Bigfoot to look at where the groupings of sightings are located and make your own conclusions. One area in California where the elevation association becomes quite apparent is in the Oroville area. There are many sightings in this same elevation region that are scattered throughout many years of sightings. I am definitely not implying that sightings and incidents don't happen at other elevations, because they do. Sightings have been reported above the tree line and many have been reported in the Lake Tahoe basin above 6,000 feet. In very general terms it would appear that Bigfoot likes moderate climates and the ability to move in and out of different zones as the weather changes. The 2,000-foot region allows the creature to drop lower when the weather gets very cold and it also allows it to step up into higher elevations, as it gets warmer. Herds of deer and elk move into the alpine regions and down into the coastal range as winter and summer come and go. Two thousand feet may be the optimum location to center Bigfoot's activity.

It rarely snows in the Hoopa basin. It routinely snows at the top of local peaks above 3,000 feet. Snow stays on the ground throughout the winter above 4,000 feet. It does appear that Bigfoot isn't intimidated by snow and regularly makes treks into it, reason unknown. It would make sense that the creature wanders out into it for hunting and gathering excursions and then returns to the comfort of its den at night, probably below snow level. The Bigfoot in British Columbia, Canada must live with snow throughout the winter in certain portions of the region. How they have accommodated themselves for sleeping and what they eat is for another book.

Affidavits & the Bigfoot Investigation

As someone who has interviewed thousands of victims and witnesses of crimes, I can attest to the importance of proper interviewing techniques. In conjunction with interviewing a victim of a crime, if you're lucky you will solve the crime and interview the suspect. Suspect interviews are much different than the victim or witness interview. Suspects tend to be evasive and sometimes confrontational. Interviewing techniques vary according to the location, background of the suspect, time since the incident, information and evidence gathered. There are also other variables.

Interviewing witnesses is entirely different than interviewing victims and suspects. The witness usually doesn't have the same vested interest in the outcome of the issue as the victim or suspect. The witness interview can sometimes reveal more information than any interview the investigator undertakes.

During the first few minutes of any witness interview I usually gauge the intelligence and educational level of the subject. I may even ask some brief questions about their past with regard to formal education. If I am not certain, I may also ask about their abilities as far as eyesight, hearing and smell. It may also be helpful to understand where someone has spent his or her youth while growing up. If they were raised in New York City versus the farmland of Iowa, this would give a witness a certain perspective. In other words, a person who spent their afternoons on a John Deere tractor might not have the same perceptions as someone who mostly had lived in a penthouse on 41st Street.

Among the thousands of witnesses I have interviewed, I have never had a witness mistake someone walking from the scene of a bank robbery as being in a wheelchair wheeling away from the scene. I also have never had a witness claim that a suspect fled a location on a motorcycle and later actually find they were driving an automobile. People who live in an urban or suburban environment simply don't make these mistakes. In the world of illicit drugs and narcotics I have never had an informant mistake a bale of marijuana for a kilo of cocaine. Narcotics informants who have lived in the world of narcotics know the difference between the two drugs and their packaging. The last comparison I will make deals with livestock. I've never met anyone associated with a farm mistake a cow for a Shetland pony. A Shetland pony is shorter in height than a normal horse, has a coat similar to a cow but its snout and facial structure is much different than any cow. Do these comparisons sound ridiculous? They are meant to, as they prove a point.

In my time investigating Bigfoot I have had many friends and neighbors question me about the validity of Bigfoot sightings. Skeptics are always challenging witnesses; it's easy to do, as witnesses are never around later to defend their sighting. The challenge that skeptics level at most witnesses is that they didn't see what they claim they saw. I think it's human nature to question, and questioning isn't a bad thing in the realm of life. However, I do believe that skeptics need to rationally analyze their questions before leveling them at an honest witness. Most Bigfoot witnesses I have encountered did not approach me; I found them. Most of these witnesses told me they were initially afraid of being publicly ridiculed and thus didn't want their story heard by the masses — until I told them of the masses of sightings that surrounded them. I have a close friend who is a CEO of a NASDAQ company who believes that he saw a UFO. He won't report it to any agency and he doesn't talk about it publicly, but he claims he saw it. Once witnesses understand that there are thousands of other witnesses to Bigfoot that are just like them, afraid to tell their story, afraid of ridicule and are adamant about what they saw, only then is the story easier to talk about. I think it is also comforting to the witness to be interviewed by someone who listens with intent, has formal education in interviewing techniques, and believes in the creature and has compassion for the witness. The witnesses also know that I will conduct a thorough investigation, and if I didn't believe their story, it wouldn't be in the book.

I've made comparisons in this section because I want the reader to understand how ridiculous some of these accusations can be. Everyone who has slight familiarity with ponies and cows will know the difference between a Shetland pony and a cow. Anyone who has spent a slight amount of time in the forests of North America or has gone through a high school biology class will know the difference between a bear and a gorilla — anyone! I've heard the skeptics claim that witnesses aren't seeing Bigfoot; they are seeing a bear. I think the same level of comparison can be drawn from the Shetland pony and the cow. I don't think a mistaken identity of that level is going to happen.

There are times in witness testimony when I hear the witness claiming they saw a creature 700 yards away on another ridgeline. The creature was moving fast and it was dark in color. The creature appeared to have hair or fur and it looked big. This would be an example of a truthful description by many campers on a weekend trip to many state parks. This example illustrates how a truthful description could be construed as a Bigfoot, bear, moose, elk, horse, almost anything on four legs.

I have heard claims that there are over 2,000 Bigfoot sightings on file in North America since 1950. In my time in Hoopa I have documented close to 50 sightings and incidents and I could only find five reported Bigfoot cases on all of the websites I searched (sightings directly on the Hoopa reservation). I have been in several areas of California where there were small numbers of Bigfoot sightings reported; yet as I spent time in the community, and made myself known, the numbers would increase tenfold. If 2,000 are on file, then I believe that we can extrapolate that 20,000 have actually been witnessed and not reported. The United States Justice Department states that only 10 percent of all rape cases are ever reported. Why aren't 100 percent reported? Well, there is public humiliation, fear of embarrassment, people not believing, etc. Maybe society, as a whole, needs to treat witnesses differently on any sensitive issue. Maybe the presence of someone who is interested, trusted and wants to document facts allows the witnesses a certain comfort zone for them to recite their story.

As I started to think through how I would document this story I wanted to develop a method for documentation and validation that was different from past authors and a method that added credence to the sighting. In thinking through legal documents, there have been many times where I have signed affidavits and declarations legally declaring, under the penalty of perjury, that the statement contained in the document is the truth. I now started to formulate my plan.

Affidavits

I've read just about every book on Bigfoot that is commercially available. In many of the books there are sighting reports with names of the witnesses excluded or altered and in some cases vague notations of the exact location of the sighting. As a customer and purchaser of the book, and someone who is attempting to learn something from the writings, I am always interested in every detail, precise locations and the veracity of the witness. I am also a little reluctant to believe everything I read. Many of the sighting reports carry fascinating information, interesting Bigfoot behavior and even more interesting witness responses. Yet through all of the reports, sightings and writings, I have never found an author, researcher or investigator to hold the witnesses accountable to their report.

As someone who has interviewed murderers, robbers, rapists, burglars, drug dealers, drug users, sexual harassment suspects, and victims and criminals from almost every other criminal venue you can imagine, the most difficult interview is with someone under the influence of a controlled substance. People who are using a mind-altering drug don't have the same physical responses when lying as someone who is sober. In my years in law enforcement I have taken many classes that taught how to distinguish an honest response versus a deceptive answer. These classes also taught the proper method to phrase a question to illicit an honest response rather than a leading question that goes towards an answer that the interviewer wants.

All of these skills were immensely helpful in my efforts to interview witnesses and their sightings, and to determine who was credible and who was not. I truly respect and appreciate the authors who have written other Bigfoot books, but I doubt that any have had the educational and vocational classes I have had in interviewing and interrogation. I believe that my background assisted me in weeding my way through the many sighting reports I accumulated.

When a witness in a criminal or civil hearing is willing to legally stand behind their claim or sighting, they may be asked to sign an affidavit or declaration. Both statements are legally binding and under both scenarios you are legally stating under the penalty of perjury that your statement is truthful and accurate. If it is found that your statement is not true, then you can be prosecuted under perjury statutes of the State of California. These statements should not be taken lightly, and they were never signed by any of my witnesses without some type of questioning by the signee.

When I first instituted the practice of requiring an affidavit, I wanted a policy that was easily understood and could be documented and tracked. The first requirement to use the affidavit was that the witness had to make a claim that they actually observed a Bigfoot. If a witness claimed they heard sounds, smelled odors or something of an equivalent nature, then an affidavit was not required. If a witness claimed an actual Bigfoot sighting, I would investigate and document their information. There were many times where I documented the interview and investigation and sent it to the witness for review. The witness would read the document, make minor changes and send the document back for correction. I would correct the document and send it back for signature. If I did not receive the document back, I would contact the witness and attempt to understand what the issue was for the delay; but this was very rare. I had several witnesses ask why I was utilizing a formal procedure (an affidavit) for documenting sightings, but I never had someone refuse to sign. There were a few occasions where the investigation results didn't support the witness statement. In those instances I would not forward an affidavit and would interview the witness again to clarify the ambiguities. If, after a second or third interview, I was still not satisfied with the veracity or honesty of the witness then the story was not included in the sightings group.

The affidavit evolved in the hopes of subtly pushing people away who may not have been totally honest. It was one more mechanism to ensure that the sighting reports we investigated were legitimate and honest, and that the people were confident in what they observed. In every case the affidavit made people think about what they were stating. Every witness read the affidavit; some changed the wording, and location or description and some asked questions. I firmly believe that all of the witnesses saw what they claimed and I doubt anyone exaggerated their incident. If the affidavit did anything positive to this investigation, it ensured that the witness statement was 100 percent accurate and there was no embellishment of the sighting.

There were incidents that I believed to be significant where I requested that the witness sign an affidavit. On those few occasions I knew I would be expending significant time and resources investigating the claim and I wanted an absolute commitment on the part of the witness to their story. I never had a witness refuse to sign an affidavit under these conditions.

A statement signed under the penalty of perjury is not a guarantee that the statement is 100 percent correct, but it is a guarantee that the person is claiming under the penalty of perjury that the statement is accurate. There are several statements in this book where there were multiple witnesses to the same sighting. In each incident all of the witnesses made an affidavit. These are compelling stories where the witness statements are extremely close in their descriptions.

There are other methods to ensure that a witness is telling the truth, but they are expensive and cynics will claim they are rigged. The polygraph test (lie detector) is an extremely accurate means to test the honesty of a sane and sober individual. It has received a bad name by the mainstream press, yet the United States government regularly uses it on its own employees to determine if they are being honest. The Federal Bureau of Investigation routinely polygraphs their own agents if there is any question they are telling the truth in an internal investigation. If they refuse to take the test, they are terminated. The polygraph is only as good as the administrator giving and reading the test and deciphering the results. To correctly and accurately administer a polygraph test and read its results, you must administer a drug and alcohol test to determine the individual isn't under the influence of something while taking the test. The total cost would be over $1,000 dollars per individual, cost prohibitive to this author.

The affidavit was one tool that was fairly inexpensive to implement, made witnesses concentrate and accurately depict their sighting, and it steered people clear of me who wanted to fabricate sightings. I actually believe that the affidavit made the witnesses feel more comfortable with my process, as it was formalized and I was legally binding the witness to the sighting; and it appeared to many to be a professional method of documentation.

The affidavit, which is a listing of the basic facts of the sighting, is a very sterile document. The feelings, expressions and color of the story are provided in my description of the incident that is in the sighting reports section of this book. Should the veracity of the sighting ever be questioned or a witness ever recants their sighting, I would refer to the affidavit for support and backing.

Chapter 4
Incidents Involving Bigfoot

Hank Masten

Foreman
Hoopa Modular Homes

Incidents #1: 1966
#4: 1970
#16: 2005

I was referred to Hank by members of his family and was told he was an individual who has had Bigfoot experiences his entire life. Hank lives approximately 1,000 feet further down Lower Mill Creek Road from Inker McCovey. Hank's residence is one of the last homes on the road before it dead-ends at Mill Creek. I made a few initial attempts to contact Hank but we kept missing each other. I finally found him home at eight o'clock on a Thursday night. Hank politely asked me to come back the next day as he was trying to get some sleep.

I arrived at Hank's house at 5:00 p.m.on Friday. It was on a lonely stretch of Mill Creek Road without neighbors nearby. Hank gave me an unnerving look and questioned me extensively regarding why I wanted to know about his Bigfoot experiences. After a 10-minute dissertation about my motives, Hank seemed to relax and slowly edged his way into his life-long experiences with the creature.

Hank started his story by saying that he is a Hoopa tribal member, graduated from Hoopa High and he has lived his entire life in and around Hoopa. He said that he is presently separated, does not have kids and is a foreman at a local Hoopa business, Hoopa Modular Homes. Hank said that the farthest he has traveled from his home is Minnesota when he vacationed with the Hoopa softball team to the Indian National Softball Finals. He said that both times that they went to the finals they lost the last game by one run.

Hank said that when he was growing up he spent a lot of time with

Hank Masten standing in the front yard where he found Bigfoot tracks. His house can be partially seen at the far right side of the photo.

aunts, uncles and cousins camping, swimming and relaxing in the rivers, creeks and lakes around Hoopa. The first unusual incident occurred when he was five years old and camping at Mill Creek Lake. The lake is an eight-mile drive outside town on a dirt road east of Hoopa and an additional short hike on a well-maintained trail. Hank said that the trout fishing is superb and the camping is quiet and serene, normally. He stated that he was five years old and camping with his parents and he remembers that it was late, dark and cold and he was in his sleeping bag. He stated that they all heard a very loud scream that seemed to be several hundred yards away. Hank said that the noise scared him a lot because it was something he had never heard. He remembers his dad telling him not to worry because it was only a mountain lion looking for a friend. Hank said he believed his dad at the time, but he never forgot the scream.

Approximately nine years after hearing that first scream, Hank was camping, netting and barbecuing on the Trinity River with some friends. They were staying on the western bank just north of Tish Tang Creek.

Hank said that it was July or August in 1970 or 1971 and it was a warm night. He and Clifton Wallace (now deceased) and three other friends had just finished setting nets in the river and had started to barbecue hot dogs and marshmallows. It was probably close to 12:30 or 1:00 a.m. when they heard a very loud scream coming from directly across the river. Hank said that he had heard that scream before, up at Mill Creek Lake when he was five years old, as previously stated. But this scream was louder than the lake incident and occurred several times over a 10-minute period.

Hank said that the scream had everyone's attention and they were a little on edge. After 10 minutes of screaming, he said that a huge boulder, over 40 pounds, came flying in their direction from the area of the screaming. The boulder landed in the middle of the river and wasn't too close, but it was close enough to scare all of the boys. Immediately after the first boulder, 6–9 more boulders came in their direction. After approximately 10 huge boulders were thrown in their direction the boys decided to get out of the area. They quickly gathered up their belongings and retreated to a friend's front lawn, safely out of boulder range.

The next morning Hank said that he and his friends went back to the river front campsite and could easily see the boulders in the sandy river bottom. He remembers all of the boys swimming out to the center of the river and trying to retrieve just one of the smaller boulders, but they couldn't move it. Hank believes the boulders weighed 40–50 pounds and were thrown in the air over 200 yards. He maintains there was no possible way that any man could have thrown the boulders the distance that they saw them propelled. Hank guessed that he wouldn't be able to get a boulder to the river's edge even if he rolled it from the flat where the boulders were tossed.

An interesting note on this encounter is that there are many documented Bigfoot incidents that involve people who are barbecuing hot dogs. The incidents sometimes involve rock throwing, yells and even visitations. There might be something about the odor, the people's presence, or maybe it is just coincidental because so many people cook hot dogs while they are in the outdoors. Maybe they could be cooking anything for the incidents to occur. Maybe Bigfoot associates the smell of cooking hot dogs with children, and this sparks its curiosity.

Hank's third incident was closer to home, more recent and an actual sighting (although classified as an incident).

The winter of 2004–2005 in Hoopa had many, many days of heavy rain. It was a tough winter that was cold, wet and dark. On a weekend night in January 2005 Hank had two female friends at his house and they were all making Indian art. Hank remembers they had stayed up the entire night talking and doing various art projects. He said it was about 4:30 a.m., raining heavily outside and very smoky in his house. He said that he needed to get some fresh air and decided to get in his car, drive to the end of Mill Creek Road and see how far the creek had risen during the night. He said that he was almost to the end of the road and had gone off the paved portion when he passed a fir tree on the right side of the car. He said that he was at the point of where he was going to turn around when he noticed a huge shadow stand up from behind the fir tree directly next to his passenger's side. He said that the shadow took up the entire right side of his car. It was close to 7–8 feet tall, but it was too dark to distinguish specific features.

Hank said that he quickly turned his car around and drove right back to his house. He parked his car, entered the house and made a decision not to say anything to the women so they wouldn't be frightened.

Hank said he was home only a few minutes and back to work on his art project when one of the women in his house looked at him and said, "Someone is outside." Hank told her that she was crazy, but she insisted that she felt the presence of someone outside. Hank said he opened his front door, quickly looked outside and closed the door. It was still dark, pouring rain, and there was no way he was going out to look; and he said he really didn't want to see anything.

At approximately 7:00 a.m., with the first light of day, Hank said he went out to look around. He walked out to the entrance of his driveway and saw huge footprints in the dirt adjacent to his front fence. The grass in the area was matted down with such force that it left a deep indentation in the ground, many inches deeper than his foot could make. He said that he followed the tracks around the side of his yard and back down into Mill Creek. Hank guessed that the tracks had a stride of over seven feet and were indented 3–4 inches in the grass and dirt. He stated that the tracks appeared to be 20–24 inches long and much, much wider than any human footprint. Hank said that the print would remind you of a barefoot human print with large distinct toes.

The next evening Hank was lying in bed and thinking about the previous night. He knew that the Bigfoot had followed him down the road from where he saw it by the creek. The question that he continually asks himself is what the Bigfoot wanted or what was it interested in; or, was it just curious?

During Hank's explanation of his incident, he stared directly into my eyes with an intense determination to explain the story methodically and carefully. He sometimes hesitated and was careful in using words that correctly explained what had happened. Near the end of our meeting, Hank volunteered to escort me into restricted areas of the reservation. He even explained how he and relatives had built roads into certain areas where tribal elders believed Bigfoot resides. The elders tell members to stay clear of these areas and give the "big people" the area they need. Hank stated that he was serious in his offer and that anytime I needed a guide into those "restricted" zones, he would take me. He also signed an affidavit covering all of the incidents and sighting.

Location

The location of Masten's sighting fits the entire regional profile of Upper and Lower Mill Creek Roads. In this incident Bigfoot never took an action toward Hank, and merely followed him back to his house. The Bigfoot visiting this area do not appear to be afraid of people, but act cautious in their approach and do not stay long in a witness's sight line. They act almost as though they are as interested in human behavior as we are in Bigfoot lifestyle.

I have spent many hours in this area and have walked many miles of the Mill Creek water line. I have seen salmon and steelhead migrating up the creek and this may be one possible reason why Bigfoot is frequenting the region. The other obvious food source is the enormous number of berry bushes. The area where Hank found the footprints in front of his house was located in a huge patch of berries. The path the creature took back to the creek was also through a huge patch of berries.

During my exploration of the region I did find one almost completely intact animal skeleton approximately 100 yards from where Hank made his sighting near the fir tree and in the region of Mill Creek behind his residence. Directly adjacent to the skeleton I also found a huge mark in the

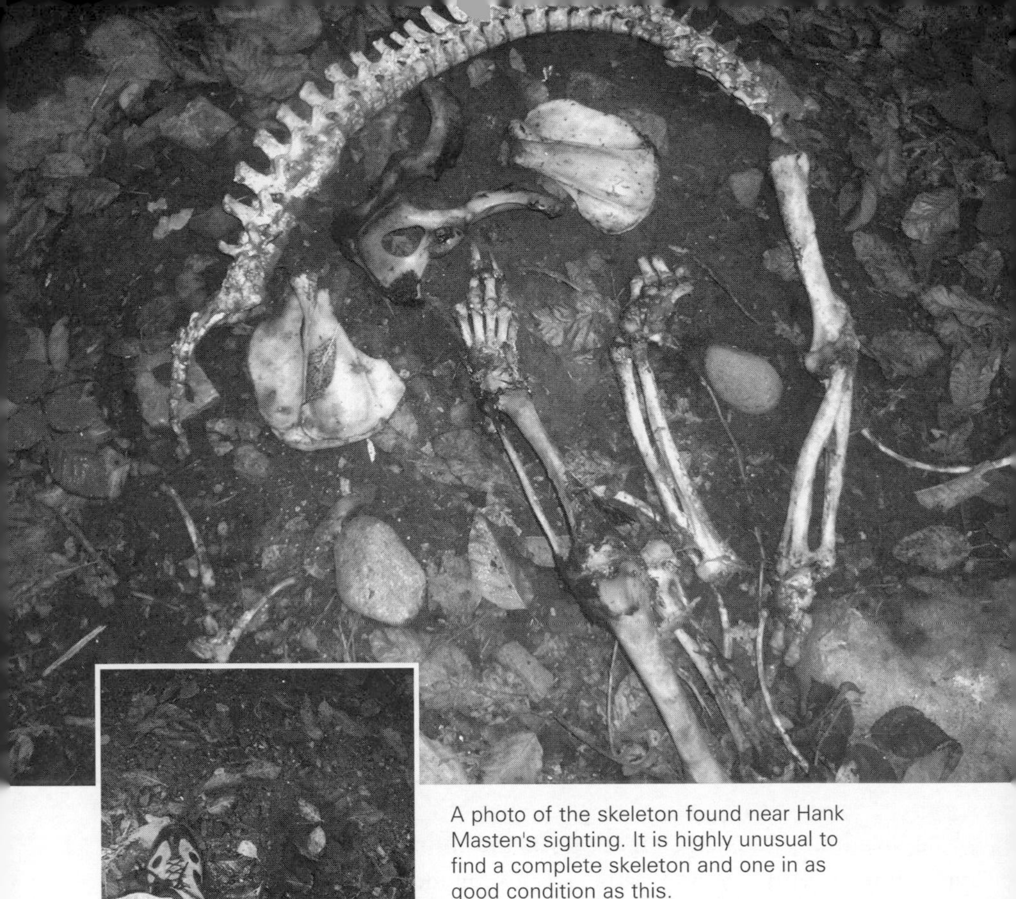

A photo of the skeleton found near Hank Masten's sighting. It is highly unusual to find a complete skeleton and one in as good condition as this.

Photo of the footprint made in water directly adjacent to a skeleton I found in Mill Creek behind Hank Masten's residence.

sand that was partially submerged in a small pool of water. The mark appeared to be a Bigfoot print, but because it was in water positive identification could not be made. I must admit that it was highly unusual to find the print adjacent to a skeleton. The skeleton had obviously been there several weeks, but the print was probably placed in the last several days. This area of the creek is frequented by locals during the summer and is used as an area to barbecue, fish and swim. During the winter months few people use this area and it is rarely visited.

I should note that Hank's story of the boulders being thrown into the Trinity River also coincides with the boulder-throwing incident involving Corky Van Pelt. The stories are almost identical and the locations are very close.

Dwight "Corky" Van Pelt

Temporarily disabled
Hoopa

Incident #2: 1967

Early one brisk winter morning I made the long drive up Bald Hills and down the lengthy dirt driveway to Corky's residence. The trip reminded me of a classic Bigfoot location, slightly foggy with patches of snow scattered on the wet ground, and very isolated. The road was filled with potholes and water and was barely manageable with any two-wheel-drive vehicle; luckily I had four-wheel drive. I actually stopped for a few minutes on my way to Corky' residence and took a few photos, as it was a memorable spot.

I eventually made my way to the residence and Corky exited the front door to greet me. He invited me in and offered me some coffee as I exchanged greetings with his kids. Corky is a big man, a Native American and a lifelong Hoopa tribal member. Corky immediately opened up and told me about growing up in Hoopa and how he was an all county football player at Hoopa High School. He said that he was a very good athlete when he was younger. He explained that he had hunted and searched the far corners of the reservation when he was a teenager and a young man, and felt that he knew the reservation better than most.

Corky said that he has held a few different jobs in his life in Hoopa. He he has been a certified timber cruiser, fire engineer and truck driver. He stated that the 48 years he has been on this earth, his back has been injured on the job many times. He constantly has pain and it has inhibited his ability to hold a job. He did say that when he was younger he was very strong and sometimes fearless.

When Corky was eight or nine years old he and a group of friends, Zek Van Pelt and Merwin Clark, went down to the Trinity River near Camel Creek. He stated that it was a fun thing to do in the summer. They would put out their fishing nets and usually catch nothing because they never did a great job placing them. Corky said that it was starting to get late when they began to hear screaming coming from the far side (east side) of the Trinity River. He stated that the screaming was like nothing he had ever heard before or since. It sounded somewhat like a creature

Corky Van Pelt at his residence.

was screaming at something that it was fighting. He explained that the screaming went on for several minutes with all of the boys being very scared. Merwin and Zek were so scared that they climbed into the cab of the truck and locked it.

Corky said that when the screaming stopped, the rocks started to be thrown. These rocks weren't small; they were really boulders, 150–200 pounds each. All of the boulders came from the eastern side of the Trinity and landed in the middle of the river. The boulders never got extremely close, but it was enough to scare the heck out of the boys. Corky said that the screams coupled with the boulders were an event he could never forget. He is positive that there isn't a man or woman alive that could scream half as loud as what he heard, and there isn't a human alive that could throw one of those boulders five feet let alone 200 feet. Corky said that he is sure that he and his friends had disturbed Bigfoot, it got angry and it was sending a message that it wanted them to leave.

Corky had one other Bigfoot encounter and this was also associated with the Camel Creek area. He stated that it occurred about 12 years past when he was hiking 5–6 miles up from the highway. He was walking for several minutes and felt that something was paralleling him on a ridge above him. He stated that he heard noises, but never saw anyone. After several minutes he decided to rest and take a short break. He said that he sat down when all of a sudden a huge bush 40 feet from him across a small clearing started to shake frantically. Corky said that this bush was much too large for any man to be able to shake that hard and fast. He said that he was scared, so he stood up and held his coat open and made large wings. He said that he wanted the creature to smell his odor and know that it wasn't wanted in this area. Immediately after opening his coat, Corky said that the shaking stopped and he heard bipedal running coming from behind the bush. The creature was so heavy that he could feel the ground shaking slightly. Corky said that he knew better than to try to chase the creature, so he just left the area.

The most recent Bigfoot-related activity occurred in the prior 2–3 months on his property where he and I were meeting. He said that he had found a dog that appeared to be half wolf. The dog was too wild to bring into the house, so he used a heavy gauge steel chain to tether it to a large tree 40 feet from his residence. Corky guessed that the dog weighed 100 pounds and was very tough. He explained that one night he and Tane (his wife) heard loud whistling coming from the area of the forest adjacent to where the wolf was chained. Both stated that there is no man they know who could pull the chain apart. Well, the next morning Corky went out to check on the wolf and found that the chain was pulled apart and the wolf was gone. The wolf has never been seen again and both Tane and Corky feel that possibly Bigfoot took it.

I should mention that the area where Corky and Tane now live is the same area where Warrior Sanchez lived and had Bigfoot encounters. This is also the same hill where numerous others, such as the Marshalls, have seen Bigfoot or had Bigfoot encounters. It's obvious that there is something very different about the Transmitter Hill/Bald Hills area and Bigfoot. There have been many credible reports from people of high integrity, people who didn't find me. I felt that they simply do not recognize that there is significant activity in this area.

The area around Camel Creek also is high on my list for Bigfoot sightings. This is around the same area where Michael Mularkey saw Bigfoot on his way into his supermarket, and also very close to the area around Shoemaker Road where Leanne Estrada and her mom saw the juvenile Bigfoot in their front yard. None of these people were interviewed together or (to my knowledge) have ever spoken about their Bigfoot sightings with each other.

Location

The location of Corky's boulder-throwing incident is less than one-quarter mile from the Hank Masten boulder throwing. The dates are approximately one year apart, yet the description of each incident mimics very similar behavior, with the witnesses conducting themselves in almost an identical manner. This occurred on the Trinity River during summer months while boys were out having fun. This really makes researchers wonder how many other times over the years identical behavior has occurred and was never reported.

Carlo Miguelena

Forestry Tech
Hoopa Tribal Forestry

Incidents #3: 1969
#13: 2006

On November 30 just after lunch I made a visit to the Hoopa Tribal Forestry office in an effort to locate a stream in the Klamath River basin. I had some new information in that area and was looking for assistance in locating a specific drainage area.

When I arrived in the office most of the staff were at lunch and there was only a receptionist at the front desk. During these types of visits I try to dress a little nicer than I do when I'm in the bush. I usually wear a pair of clean jeans and my khaki long sleeve shirt with the "California Bigfoot

Search" logo. I always have my notebook in my hand and business cards in my wallet. When I announced to the receptionist who I was and what I represented, she appeared a little shocked. She did say that there was one person in the back who might be able to assist me.

The receptionist walked me to a back room and I was introduced to an individual from the forestry group. He was in his late thirties, professional and well spoken. He stated that he could show me on a map the location I was looking for and then asked a series of interesting questions. This person had spent his entire life in the Hoopa area and truly didn't believe in Bigfoot. He explained that he had spent countless hours in the most remote regions of the reservation and he had never seen any sign of the creature. He stated that he has heard rumors of people who have seen Bigfoot, but doubts they have truly seen something similar to what they are claiming. Just as this individual was letting me know his opinion, another forestry employee heard our conversation and started to level the same opinion my way. Both individuals were polite and professional, but true pessimists on Bigfoot.

After I was shown the location for which I was searching on the map, I was walked to the lobby and more employees from forestry and also wildlife management started to enter the building. Within a few minutes there were almost 10 people in the lobby area who were all talking about Bigfoot. Once the wildlife management staff arrived, the tide of opinion started to change. They seemed to be a much more optimistic group when it came to Bigfoot and its possible existence in Hoopa. Once the opinion in the lobby started to change, one of the forestry employees, Carlo Miguelena, stepped forward to make a personal statement. Carlo had the strength to speak his feelings and opinions in the face of obvious apprehension. He stated that he had a Bigfoot encounter, and the entire Hoopa forestry and wildlife room then went silent.

Carlo explained that he is a 53-year-old forestry tech for Hoopa Tribal Forestry. He stated that he grew up in the valley and spent his life in the mountains, streams and creeks surrounding Hoopa. He graduated from Hoopa High School and has heard stories about Bigfoot his entire life. He explained that when he was young, the Trinity Alps National Wilderness Area had not been established and where the tribal grounds stopped on the eastern edge, national forest began. He said that when he was young his family raised 25–30 cattle on forest service property just outside the reservation near Water Dog Lake. His family consisted of his mom and dad and an older brother and younger sister.

Every summer the family would spend at least a week near Water Dog and stayed at the Trinity Summit Lookout cabin. Carlo said that during his life his parents never got into long discussions about Bigfoot and never discussed their beliefs with him. In the summer of 1969 Carlo was 15 and his family was taking their annual trip up to Water Dog. He stated that he can distinctly remember everyone loading up the family's Jeep Wagoneer and making the drive up into the hills. He remembers that they traveled up Big Hill Road until it turned to dirt, and then continued at a slower pace as they gradually went up in elevation. They were approximately 12 miles from the state highway and at 5,700 feet when they hit a very rough and steep location in the road. This area is now in the wilderness area and the public cannot drive it, but he remembers how rough it was and how slow they were traveling. Carlo also remembers that it was a very hot day and because they were in a steep section, his dad had the air conditioning off and the family had all of their windows open.

Carlo Miguelena in front of the Hoopa Tribal Forestry office.

Miguelena said that at the point his family's car was driving slow and as they were ascending the large hill an incident happened that he would never forget. He said that the entire family heard the loudest scream he has ever heard in his life. He explained that he has heard mountain lion screams many, many times and knows precisely what they sound like. Carlo stated emphatically that this scream was not a mountain lion or any other animal in Hoopa that is accepted by scientists as a member of the animal kingdom. His entire family started to look at each other when the scream was heard. He said it was hard to tell how far away the screamer was, but he was sure that it wasn't far away and that it was probably sending his family a warning.

It was either Carlo's brother or sister who asked the parents what was making the scream. Carlo said that his dad immediately replied, "Bigfoot."

It wasn't a quarter of a mile and Carlo said that his dad had turned their Jeep around and they were heading home. There was no family discussion and there was no dialogue about the decision. The Miguelenas were heading back to Hoopa, period.

In the 53 years Carlo has lived in the Hoopa region, he's had only one Bigfoot encounter. Carlo said it has only taken one direct encounter to convince him that something large, loud and menacing lives in the forests surrounding Hoopa. He said that he has heard many friends tell stories about their sightings and their beliefs and he thinks many of their stories are quite credible. His best guess is that Bigfoot lives in the far reaches of Tish Tang, possibly into the wilderness area to avoid public contact. Carlo stated that he has heard too many stories, knows there is plenty of food and cover to maintain a population and hide something the size of Bigfoot in the outer reaches of the reservation and beyond.

Carlo said that he would gladly feed me any leads and provide any new information should it develop. After we had a lengthy discussion, there were 4–5 additional employees who came to my vehicle and had private conversations about Bigfoot. One of the greatest Bigfoot stories I have ever heard, and is chronicled in this book, came directly from the family of one of these employees. I sincerely appreciated their inner strength and their belief in my integrity. Thanks guys for having the belief in my credibility and stepping up to be interviewed. The world appreciates your stories.

Update

In mid March of 2007 I was in Hoopa on one of my regular trips when I saw Carlo at the gas station. The station is a general meeting spot in Hoopa and if you are there long enough you will eventually see everyone, as it's the only place in the area for gas. The tribe owns the station and keeps prices comparably low for tribal members, yet everyone gets the discount.

I was filling my Jeep when Carlo came up to me and asked how the search had been going. After a short conversation Carlo said that he had recently seen Bigfoot tracks on the backside of Big Hill. He stated that he and his son had driven into the backcountry on some free time. He explicitly stated that they had gone way too far and driven into several miles of virgin snow and could have been stuck because the snow was very deep where they were. They exited their vehicle and walked around and saw one set of huge footprints. He said it was obvious that they were the only vehicle that had been in the area for weeks, and the prints were fairly fresh and an obvious Bigfoot track. He stated that the stride between the prints was much larger than a human stride, and he estimated that the prints themselves were over 16 inches. He also said that he could tell they were made by a bare foot and not a boot. Carlo told me that he has seen hundreds of bear tracks in his life and is positive that the tracks they saw were Bigfoot and not any other natural wildlife.

Several days after Carlo told me of his footprint sighting, I drove back into the region. I also encountered heavy snow with many fallen trees. The winter of 2007 had been very tough on the Hoopa reservation. Ed Masten had told me that every 10 years the oaks and other hardwoods usually suffered a similar termination of life through heavy winds and snow. It would appear that 2007 wiped out many good trees on the reservation. My trip into the region that Carlo had visited produced no sightings, but a lot of intelligence. It was much more desolate than other regions of the reservation and it was obvious that nobody should venture into this area without four-wheel drive and great tires. I meandered around many downed trees and lots of huge mud puddles. I eventually reached an area that had several feet of snow and I felt it was time to turn around. When I reach a point when I know I am turning around I usually exit my vehicle and walk several hundred yards further down the road just as an insurance measure that I'm not missing anything that's close. I did this at this location and didn't see anything unusual. I will be back to this spot over the spring. According to topographic maps, I was on the eastern fringe of the reservation and in United States Forest Service Property.

Jeff Lindsay

Hoopa Forest Planner
Hoopa Tribal Forestry

Incidents #5: June 1976
#6: July 1980

The Hoopa tribe requires a special permit for any non-tribal members to be on reservation property. Anyone can be on the paved section of the reservation but once you travel onto the dirt roads, you need a permit. The procedure calls for the Hoopa Tribal Forestry Department to review your request, understand the need and then inspect your vehicle for contaminants. There are several nasty tree diseases that exist in Northern California and the tribe doesn't want any of those contaminating their trees. The tribe also doesn't want people harvesting their trees or stealing wild mushrooms. They want their forests to stay pristine, and they see the inspection as a key to keeping forests pure. They require that your vehicle be absolutely clean of any dirt and brush before they will issue the permit. The permits are for a maximum of one week and then you must go back through the same procedure. The person assigned to inspecting vehicles is Jeff Lindsay.

It was early one Wednesday morning in March when I first met Jeff. He is a 44-year-old Humboldt State University biology graduate who greeted me with a big smile. I explained that I was there for the vehicle inspection and that I promised the car was clean. He looked happy. We walked out to the car and he looked it over carefully. Jeff said it was quite clean and it passed. It was a rainy day and we went back under the shelter to talk.

Jeff asked me politely and specifically what I was doing in the woods and what success I had experienced. Jeff's questions started a healthy

Jeff Lindsay at his desk in the Hoopa Tribal Forestry office.

conversation that usually always ends with me asking the person if they had Bigfoot experiences. Jeff said approximately 15 years ago he did have a Bigfoot experience while in the Trinity Alps. It was now clear why he had curiosity about my exploration.

Jeff said that he grew up in Lodi, California and belonged to a Boy Scout troop that was led by his dad. He said that his dad always had an interest in the Trinity Alps and they seemed to go there more than other places. He stated that it was in June of 1976 when he, the troop and his dad backpacked into Papoose Lake, deep in the Trinity Alps.

When Jeff stated "Papoose Lake" I immediately remembered a sighting report I had read a year earlier about a Bigfoot incident at the same lake. In that incident, a group of campers were scared out of their mind by screams, yells and trampling of the ground around their tents during one stormy night at the lake. They never saw what was causing the disruption, but they claimed it was Bigfoot. The campers were so scared that they said they would never go back to the lake.

Jeff stated that there was still a lot of snow around the lake and its surrounding basin, and at the time they decided to explore the area. He immediately found a set of prints with a very long stride and huge footprint, much larger than a man's and the impression of a large bare foot. He stated that the prints went up to a ridgeline at which point he decided to stop. Jeff walked back around the bowl and up another route and found the same prints apparently meeting up with two other creatures where they appeared to be all walking together. He stated that all of the prints went towards a very large crevasse, large enough to walk through. He stated that he and his friends walked to the opening of the crevasse and then got very scared. He said that a feeling came over him, like he was being watched. Jeff said that something inside told him that he shouldn't be going in there. Jeff claimed that he and his friends got out of the area after seeing the huge opening. He said that he had a camera and took a picture of the prints but the photo had too much white in it from the snow and didn't show the detail from the print.

Jeff said he had one additional encounter that he attributed to Bigfoot and that also occurred in the Trinity Alps in early July of 1980. He was with his dad and the troop when they were on the trail to Papoose Lake. He said that it was a two-day hike to the point they were at — Enni Camp. They had stopped to take a break and casually walk the area without their packs. He said that as they were starting to walk around, a couple of teenagers began to walk towards them from a creek bed that was nearby. The teenagers told them that there was too much snow at Papoose Lake, so his dad decided that they would camp at Enni for the night.

Jeff said that he decided to slowly walk the creek bed that the teenagers had just walked down. He was looking at the creek and soil and had walked approximately 100 yards when he came upon a huge footprint in the lone sandbar, the only one that was anywhere nearby. Jeff said that a boot print (obviously from the kids that had just walked down the creek) was near the print. The footprint was between 13 and 15 inches long, had five toes and was much deeper in the sand than the boot print. Jeff said that this print closely matched the prints he had seen in the snow at Papoose Lake years earlier. Jeff feels that it must have been Bigfoot that made this print.

There are several aspects of this incident that are very interesting in the Hoopa Bigfoot scenario. Remember that many people believe that Bigfoot lives in this wilderness area and travels outside for food source.

It is also fascinating that a Bigfoot incident report already exists for Papoose Lake and is under different circumstances. I spoke to Jeff about this and he stated that he had never filed a report before meeting me. The final notation to remember is that Jeff described the prints he found by the lake had been on a ridgeline; this fits with a Yurok belief that the creature likes to travels along ridges.

David Bishop

Prospector
Orleans

Incident #7: August 1980

During one of my weeks spent at Hoopa I was checking into the local hotel when I was given a message to contact Dave Bishop. I'd never heard of Dave and there was no explanation to the note. I had a busy week scheduled, but I did call him the first full night I was in town.

Dave was a very congenial guy on the phone and stated that he didn't have a face-to-face confrontation with Bigfoot, but knew from his history as a mountain man and prospector that he had a close encounter with Mr. Bigfoot. Dave stated that he lived in Orleans and asked if I would be in the area anytime soon. We made arrangements to meet at the restaurant in his town later in the week.

Dave walked into the restaurant and he immediately struck me as a seasoned mountain man. He was a very straightforward guy who wasn't looking for publicity, because he made it clear from the beginning that he never saw the creature and was never claiming anything sensational in his story.

Dave stated that he was a 58-year-old prospector who had lived in and around Orleans for over 30 years. He said that he had two bachelor degrees from Humboldt State University and for a good portion of his life he was a professional student. He said that it was obvious he was going to have to work for a living and so he decided to be a teacher. He did this for a few years and got tired of the classroom. He said that he had spent time in and around Orleans as a student and liked the remote aspect of the city and truly enjoyed the prospecting it offered.

Dave Bishop at the coffee shop in Orleans.

Dave said that he and his wife had taken many daily trips in the area immediately adjacent to Orleans searching for possible mining locations. On one of these trips he found a location that appeared to have real prospecting possibilities. In August 1980 he and his wife set up camp six miles up Red Cap Creek from the confluence of the Klamath River. They didn't have much money, so they lived in a fairly frugal manner, small shelter, few provisions and a makeshift outhouse. He said that they slept in a teepee with their blankets on the ground, and dug a trench away from their sleeping quarters where they went to the bathroom. They cooked on an open fire just outside their tent and made no particular efforts to discard their leftovers. Dave did say that he and his wife often ate fish that came from the creek, and he attempted to live off the land when practical.

Mr. Bishop claimed that they were at the mine site only a few days when his wife woke him at 2:00 or 3:00 a.m. saying that there was someone (not something) outside their tent. Dave stated that as he was lying on the ground he could feel the ground shaking as though something huge was making the ground rumble. He said that they could feel something that was bipedal moving around their campsite and it made him very nervous. He said that he didn't want to get his wife nervous because they'd only been there a few days and he liked the existence they were living; he didn't want to leave. He said that he did the "manly thing" and looked out the teepee door and didn't see, hear or smell anything. He later stated that he really didn't want to see anything and really didn't make a huge effort. Dave said that the shaking of the ground and footsteps quickly subsided, but he didn't sleep at all that night. To the best of his memory there was no wind and there was almost no moon.

The next morning Dave was up at the first sunlight. He did find a partial footprint that looked a lot like a bare human footprint, but it was huge compared to a man's. He said that it was only the front portion of the foot, which included the toes. Dave can remember that the toes were huge and there were only four he could see. The ground in this area according to Dave was not very conducive to tracks because it was so hard; he felt lucky to find the one print.

While scouring the camping area for evidence of the visitor, Dave said he went to his trench toilet. Inside the trench he found a huge pile of scat, the biggest he had ever seen in his life. Dave guessed the scat was as big in diameter as a beer can and as long as a six-pack. The scat was very fresh and contained ingredients that were consistent with a bear, but it was much too large for any bear scat he had ever seen or could ever imagine. Dave said that we all know that bears do not leave their scat in trenches; they leave it on trails, roads, fields, etc. It was almost as though the creature that left the scat was trying to mimic the behavior of the people who left their scat in the trench.

Dave Bishop has owned property up Red Cap Creek for over twenty years. He stated that during that time he has seen the population of deer decline considerably while the population of black bears increases at the rate the deer are decreasing. Dave said he is absolutely convinced that a Bigfoot, not a bear, visited his camp up Red Cap Creek. He said that he has never had a similar encounter even though he has seen a large number of bears in that same period of time and in the same general area. Dave stated that bears don't shake the ground when they walk, and he signed an affidavit covering all aspects of this incident.

Location

Red Cap Creek is famous in Bigfoot history, dating all the way back into the early 1960s. Roger Patterson and his team had made many trips up into the headwaters of Red Cap. At one point they felt they had found some type of Bigfoot bed up from the creek, overlooking it on a small hill. They worked the area for many weeks, but found no other evidence. Dave Bishop never made any mention of this or any other Bigfoot activity in this area.

Red Cap Creek is very close to Bluff Creek, Klamath River and the Butte region that Hoopa elders have told their people to stay away from. Red Cap is also on United States Forest Service land, but adjacent to the Hoopa reservation. Red Cap is located on the opposite side of the Klamath River from Bluff Creek, but it is all within a short distance from each landmark described above.

Dave's comment about an abundance of bear and lack of deer makes a lot of sense, as it matches comments made by Elaine Creel from Hoopa Forestry and Wildlife. In the many weeks I have spent in this region, it is extremely rare that I see a deer. On the opposite end of that scale, it is a rare day that I don't see a bear. One possible reason for the lack of deer is the number of Native Americans hunting them year round. This argument makes some sense, but the Natives can't hunt on United States Forest Service land unless it's hunting season and they have the proper state permits, and still you don't see many deer on their property. As a general rule, the Natives don't hunt bear as they consider them too close to man. I've had more than one Hoopa friend tell me that when you see a bear stretched out and drying after a kill it looks a lot like a man.

For Incident #8 refer to Incident #2, page 99
For Incident #9 refer to Sighting #11, page 215

Wendell "Winkle" White

Retired Fisherman
Hoopa

Incident #10: 2004

After interviewing witness Debbie Carpenter, she took me to meet her husband, Wendell "Winkle" White. She had told me that Winkle had some strange Bigfoot experiences that he'd like to explain to me.

Winkle joined us during the end of Debbie's interview and explained that he is also a Hoopa tribal member and that he and Debbie had recently married. He said that he was raised with Bigfoot stories and knew that

The location on the Trinity River where Winkle White heard the screams and saw large footprints. Note the vertical break in the rock on the opposite bank where there is a dirt path between the crack in the rocks. Tribal nets are in the water attempting to catch salmon. This is also an area where two ridgelines meet from opposing mountains, and is also one of the few spots south of Weitchpec where there are no cliffs protecting the river, thus making a crossing easy.

the creature frequented the area of Debbie's house. Winkle stated that he has been down to the Trinity near the fishing grounds of Debbie's father (inside Bigfoot Alley) and has heard the creature's yells and has seen footprints along and around the bank of the river. He volunteered to take me to the spot where he made the observations.

I followed Debbie to her house and she dropped Winkle with me, and he and I drove to the river. Approximately one-quarter mile from Debbie's residence we left the paved roadway and took a very rough dirt road approximately half a mile down to the Trinity. The dirt road was fronted on both sides with nothing but huge berry bushes. There were large piles of bear scat approximately every 100 yards. After a short trip, the road opened up into a beautiful river frontage that had a monumental sized rock that added an unusual touch to this section of the river. This section was wide and deep and had an obvious opening on the opposite bank where, between other rocks, a pathway down to the water could clearly be seen.

Winkle told me that, days after Debbie made her Bigfoot sighting, he was at this location when he heard loud screams or yells coming from the forest edge near the opposite bank (up Transmitter Hill). He stated that he has lived in the woods his entire life and he had never heard anything as loud, as long in duration or as different as the sounds he heard. He also observed a series of human-like bare footprints on the bank near the water. He stated that the prints were huge, nearing 16 inches in length and much too long and wide to be human, and they were indented too deep into the ground for a 200-pound male to make. Winkle said he is positive that the tracks he saw were Bigfoot prints and not bear or human.

Location

The location of the prints found by Winkle is very significant. This area is between the Mill Creek Road subdivision and Bald/Transmitter Hills, both areas with Bigfoot activity. Mill Creek flows into the Trinity River a few hundred yards south of this location and the activity on Bald Hills Road is just across the river and up the hill from the footprint location. Debbie's sighting is in the middle of a large flat area of Hoopa that has a huge number of berry bushes. In fact, Debbie saw Bigfoot standing next to berry bushes. The creature later jumped over the bushes and walked parallel to other bushes in an adjoining field. This area of the Trinity is easily identified by several huge rock formations. These rocks can be identified from miles up Bald Hills and Mill Creek, which may be one reason that it appears to be a gathering place for Bigfoot. This is also a location where the river makes a large turn and starts to head straight for Weitchpec. The public should be warned that there is no legal access to the riverbank in this area, and you must possess a tribal permit for access.

Doreen Marshall

Yurok Tribal Member & College Student
California State University at Humboldt

Incident #12: May 2005

While I was visiting Inker McCovey at his house, he told me that a girlfriend of one of his sons had also seen something that appeared to be a possible juvenile Bigfoot. He said the girl was at his house and would talk to me.

Doreen Marshall came into the family room. She is a short, very attractive young woman who spoke very directly and succinctly. She stated that she had just graduated from the local high school and has been dating one of Inker's sons. She said that she has always been a pessimist about people who see Bigfoot, or even elders who claim to have seen "little people." (Little people are part of the tribal heritage. They come out at night, are not traditional in appearance and it is very rare to see one.) The observation made by Doreen is quite odd; however, it might have been a young Bigfoot.

Doreen said that late the previous May, just prior to finishing school, she was driving home alone on Highway 96 from the area of town. She stated that she was turning from the highway onto Mill Creek Road, near the area of other sightings. She said that she was just down the road a few hundred yards when she saw something near the side of the road. She said that the road turned slightly and the headlights illuminated a small creature, maybe three and a half feet tall. The creature had an unusual head, almost deformed, or buffalo or horse-like. She said the creature was small and had wide shoulders. It was turned slightly away from her but turned when the lights illuminated it and she could see it had dark eyes. She said it started to move very quickly when it saw her and ran, on two legs, towards the creek bed. She went directly to Inker's house and told her boyfriend. The boyfriend and family were there during the interview and confirmed how shaken she was when she arrived at their residence.

Doreen stated that she isn't sure if she saw a young Bigfoot — but it had hair over its entire body and ran on two feet — or if she saw something that elders describe as a little person. Doreen said that she thinks about the sighting every time she drives by that location.

Forensic Sketch

I contacted Doreen by phone and requested that she meet with Harvey Pratt, my artist, and see what they could sketch. Initially she stated that she would meet with Harvey, however, after several phone calls the meeting did not materialize.

I included Doreen's statement and descriptions in this document even though Doreen never signed an affidavit. She felt that this creature might be a little person and not a Bigfoot, and she wasn't positive about what she observed.

This story is listed as an incident because Doreen cannot confirm significant details to determine what type of creature she saw. Many of the details are similar to a Bigfoot sighting, but there are enough issues with the description that make it slightly vague. She obviously saw something that night that was not in the "routine" sighting mode for animals in the Hoopa area. I know that Doreen and I aren't sure what she saw.

Many months after meeting Doreen, I learned that her father, Lyle, is the tribal chairperson, something she never mentioned.

John Harlan

Professional Photographer
Orleans

Incident #14: 2005

During my initial visits to Hoopa, Weitchpec and Orleans I routinely drove through public areas and looked for government employees who were in the area. These people were generally reliable and could lead me in the right direction to obtain information on topics of interest. Several locals told me to travel to Fish Lake, which is in the area between the Klamath River and Bluff Creek. There are several sightings of notoriety that talk about Fish Lake. I spent approximately six hours in and around the area and found it ideal Bigfoot habitat. It is approximately nine miles up to the lake from Highway 96 and it is a very winding road through very dense forests. I walked the perimeter of the lake and even sat and fished, watching. The forest around the lake is quite lush and thick. All of the creeks in the area flow into the Bluff Creek basin. Fish Lake is approximately 1,200 feet above the Klamath and the mouth with Bluff Creek. The day I was there I saw falcons, hawks, squirrels, and caught a lot of trout; in fact, I caught my limit that day. There were only a few campers who stayed the night, and all were gone when I drove through the campgrounds. The area is very quiet, serene and sometimes I actually got the feeling that I was being watched — a bit strange. Near the end of my visit at the lake I met a park service attendant who was cleaning the area and asked him about Bigfoot. The attendant referred me to John Harlan, a park live-in attendant (host) at a local campground.

I arrived at the campground and was immediately greeted by John. He has been a volunteer at several campgrounds over the years; he knows the local folklore and has contacts in the community. I explained to John that I had just met a ranger at Fish Lake who referred me to him. John's face immediately lit up and he stated he knew a lot about Bigfoot. John stated that the Fish Lake area is in the middle of the Bigfoot activity for this region. He claimed that Bigfoot comes out in the middle of the night, 3:00–3:30 a.m. and wanders the campground. He has heard stories that Bigfoot can see infrared light and thus lights and night vision don't work with him. He said that he heard several stories in the previous year that Bigfoot yelled in the middle of the night and scared campers. He recommended that I sleep all day and stay up all night and have the experience. Uh, okay.

John stated that in the last year he had heard Bigfoot screams in his campground. He said that the screams are like nothing you can imagine and are nothing close to anything we normally hear in the wild. He stated that when people start coming to the Klamath area in the summer, Bigfoot sightings become scarce.

John told me that late the summer before an occupant in his campground came to him to report that he found a Bigfoot print at the mouth of Aikens Creek and Klamath River. He said that this is only a few miles from his location and it substantiates the sounds he has heard in his area. He also stated that Bluff Creek (location of famous Patterson–Gimlin film footage) is only down the road and still is a haven for Bigfoot seekers.

It should be noted that John's statement about hearing screams at night coincide with the statements of Phil Smith at the Bluff Creek Resort, which is just a short drive down Highway 96. His statement about Fish Lake being at the center of Bigfoot activity for this region is also believable. A search of sighting reports has a few incidents noted at the Fish Lake campgrounds. It is almost a straight hike from the back of Fish Lake to the Bluff Creek area. It was quite interesting to find that all hiking trails on the back of Fish Lake to Bluff Creek were closed by the Forest Service because of some type of tree disease, an unusual coincidence.

Manual "Warrior" Sanchez

Director, Qosos Networking Project

Incident #15: 2005 (two incidents inclusive)

Inker McCovey took me to the Hoopa newspaper office where he knew of someone who had another strange Bigfoot encounter, Warrior Sanchez. Inker made the introductions and Warrior immediately greeted me warmly. Warrior stated that he was in the middle of working on business regarding the Qosos Networking Project, but he would gladly give me an extensive interview and take me to his residence if I could come back at 11:00 a.m. I agreed to meet him back at his office in two hours.

I met Warrior and he asked if I'd follow him up to his residence. He got into his vehicle and I followed him the five miles to his house. Warrior lives on Transmitter hill on the opposite side of the valley from Inker. The hill is named "Transmitter" because the tribe has a huge transmitting tower located in plain view on the side of the hillside. We drove along the base of the valley in a northerly direction and then went up Transmitter Hill Road to a location near the top of the mountain where the transmitting tower is based. We were at approximately 2,700 feet in elevation.

We left the paved portion of the roadway and made our way on a very rough dirt road. We passed approximately four residences until we reached a steel-framed gate. The gate was open and we continued down the rough road for another mile. We arrived at a small clearing, about two acres in size. A small residence that appeared to be built up off the ground was in the center of the clearing. The residence faced downhill and towards a large forest area. It had a small porch that faced the forest.

Warrior Sanchez residence.

As I got out of my vehicle I saw two chickens running through the yard, and I watched as Warrior exited his car and was greeted by a large dog, a mix of Labrador retriever and rottweiler. Warrior introduced me to his wife, who was pregnant and standing on their deck. Warrior also introduced me to his dog and explained that it was fearsome and had taken on many bears in the last four months. Warrior showed me some open scars that the dog had obtained from the bears. The dog was scary looking!

Warrior immediately pointed to the woods in front of his residence and explained that this is where Bigfoot lives. He said that this is where the sounds, rocks and branches breaking come from. The area was not very thick; you could see several hundred yards through the forest. There was little shrubbery on the forest floor, but it was covered in thick pine needles and decaying leaves.

Warrior stated that he and his wife moved into their residence approximately four months earlier. He also casually said that his dad is one of only two remaining medicine men in the Hoopa tribe. He stated that he has lived in the Hoopa valley his entire life and is extremely familiar with local wildlife, their sounds, habitat and behavior. He said that the residence that he is renting had stayed vacant for over five years with nobody visiting or maintaining the area. He told me that I could see from our drive that the area is very remote, with scarce human activity.

Approximately one month after the Sanchezes moved in, they noticed that their chickens were disappearing. They were well aware that mountain lions are in the area, and initially attributed the loss to a natural predator.

Warrior stated that 2–3 months before, just after dusk, he and his wife heard large branches breaking in the forest area in the front of his residence. He said that it was just starting to get dark and he decided to grab his .22-caliber rifle and scare the bear that was making the sounds. Warrior said that he made his way to his porch when he saw a huge figure that appeared to be standing erect at the edge of his forest. He said that he purposely shot high over the head of the creature in an effort to scare it. He said that he and his wife immediately heard an extremely loud prehistoric sounding yell that was piercing in its intensity. Warrior said that he has never heard that sound before and knows it doesn't come from a bear, wolf, coyote, elk, mountain lion or any other large mammal that he is familiar seeing in the woods.

The yell lasted several seconds and was followed with the creature

Warrior with his dog at the forest line pointing to the area where he shot at the creature.

leaving the area and several large branches breaking. Warrior said that he turned to see his wife's expression and noticed his dog sitting at his feet. He was shocked to see the dog sitting on the porch and not chasing after the creature. Warrior said that he slightly nudged the dog and said, "Go get 'em." But the dog wouldn't move from his feet and kept his head lying on the floor, almost cowering. He said his wife had a shocked look on her face and said, "What was that?"

The night after the yelling and shooting, Warrior claimed that he and his wife were lying in bed when they heard a large rock hit their bedroom window. He reminded me how isolated his residence was from any neighbors and also that there are no animals that throw rocks. I told Warrior that it has been reported that Bigfoot has been known to throw rocks at people in their residence, on trails and while camping. Warrior looked shocked.

The latest incident occurred two days prior to my arrival at their res-

Location of first chicken feather site.

idence. He said that it was again approximately 9:00 p.m. and had just gotten dark. He heard a loud, deep coughing type sound, almost a "woof, woof," but very, very deep and loud. He said that he went out to his porch and his dog was again lying at his feet and wouldn't move. He tried to prompt the dog to go after the animal, but again it wouldn't move. Warrior said that he stayed out on the porch listening and watching for another 30 minutes when he heard large branches breaking in a manner that indicated the creature was leaving the area.

As I talked with Warrior, he agreed that the behavior of the creature up to this point was mild intimidation. He said that he and his wife were scared, but they weren't going to leave. He also stated that he had never had the time to walk down the hillside and check out the forest where the sounds and the Bigfoot activity have occurred. I asked Warrior if I could investigate the forest and the mountain. He said, "Sure."

As we made our way down the hillside, the thickness of the forest and the type of vegetation strongly reminded me of the forest near Inker's residence. We were approximately 30 feet into the forest and at an angle where Warrior's windows on his residence could not see the location; I found a clump of chicken feathers. Warrior immediately stated, "There's my chicken feathers, but where is the chicken?" It was a unique setting because it appears that something plucked the feathers off the chicken but there was no carcass or blood.

Warrior and I continued further down into the forest when we found a second site containing chicken feathers. This site appeared to be older than the first and had more feathers. It was obviously another location where something plucked the chicken. This was an odd site, as predators usually kill and eat their prey on the spot.

Bigfoot Den/Shelter

As Warrior and I walked further into the forested area, I noticed a grouping of approximately 20 trees that seemed to be bent down towards the ground. Most of these trees appeared to be spruce, with an approximate diameter of 4–6 inches and ranging in height from 70 to 90 feet. The trees were all bent over with their tops meeting at a common location. All of the trees seemed to be rooted in a circular fashion and purposely bent towards the center. None of the trees appeared to have had their trunks perforated or broken, and all appeared to have leaves that were still alive

Opposite page, top A long-range photo of the trees bent, not broken. The nest/den is located near the tops of the trees on the ground.

Opposite page, middle A good view of how trees from all angles come together to form a canopy near the ground. Note how all of the trees have leaves at their tips. None of the trees are dying.

Opposite page, bottom A closer photo of the tops of the trees as they start to come together to form the nest. The nest can be seen at the far right of the photo.

near their tops, which were near the ground. Some logs that were 10–15 feet long, and larger in diameter than the spruce held down the treetops. Each of the trees that was bent to the ground was carefully overlapped, one holding another until you reached the larger log. There were several other logs neatly placed in the area, which offered some insulation around the main centered area. The area of this den was the only flat piece of land on the hillside, yet offered an overview of everything below it. It would be the perfect location to build a small shelter.

The trees that were pulled down and held into the common center area could not have been pulled down by any man. These trees are much too thick, heavy and bulky for any one person to pull them down. If a few people were going to attempt this they would need to be on a 30-foot-tall crane to get the leverage needed to pull them down. Once they were down, it would take tremendous force to hold them in place while you maneuvered something into the center to hold them down. I examined several trees for signs of chains, ropes, scratches or marks that would indicate what was used to pull them down, but found none. The center area of these trees would offer a great nesting area and protection for an animal. The ground area was very soft and thick with leaves and decaying material. There was no chance of finding any footprints in this area due to the lush ground cover of leaves. The area under the trees was very matted, as though something had been bedding there.

Warrior reminded me that nobody had lived in this area for over five years. Once he moved into the area he started to be the victim of odd occurrences, yelling, howling, rocks, and whooping sounds. It would appear that Bigfoot had made his home in this area for some period of time that the land stood vacant. The den that we discovered could not have been made by any other mammal living in this area other than a Bigfoot. It is also very odd that all of these trees are still alive and hadn't suffered a break in their limbs or trunk. Warrior signed an affidavit covering all aspects of the incidents that occurred.

Location

Warrior's residence is sitting on an eastern facing slope on Transmitter Hill Mountain. This area is also known as transmitter hill for the large transmission tower that is located adjacent to Warrior's house. The Trinity River is located approximately 2,300 feet down the mountain and the Klamath River is located 10 miles to the north. This area is covered with berry bushes and bear scat. If you drew a straight line from Warrior's residence, down the hillside, across the river you would end up in the area where Debbie Carpenter had her sighting. If you continued on that line you would get into the Mill Creek Road area where there have been several sightings. Once all of the sightings have been mapped, and you can get an overall view of the area and history of sightings, these observations all start to fit into a profile. This is also just up the hill from where the Marshalls had an incident at their residence and a few miles away from where Jackie Martin saw a Bigfoot cross the road. This location is in the middle of many sightings and incidents. This location is in the western area of Bigfoot Alley.

Christy Brown and James Marshall

Hoopa

Incident #16: 2005–2006

There are times in any investigation when you must canvass a neighborhood for clues, evidence and witnesses. Bigfoot investigations are no different than any other investigation; you must commit all of your energy to answering the questions, finding the people who can shed light on evasive answers, and use your analytical skills to convince people there is an answer to the puzzle.

In most investigations the canvassing of a neighborhood includes walking door to door, knocking on neighbors residences and asking typical questions, who, what, where, when, why and how? Bigfoot investigations are just slightly different when it comes to canvassing neighborhoods. In most Bigfoot cases I have researched, there are no neighborhoods; people in those areas usually don't have neighbors close by, as was the case with Christy Brown and James Marshall.

I had been working in Hoopa for almost a year when I decided I needed to find additional information and witnesses on Transmitter Hill. This is the mountain that sits at the far northwest side of the Hoopa Valley on the west side of the Trinity River. This is the same hillside where I met Warrior Sanchez and investigated his family's encounters with a possible Bigfoot.

I spent two days driving the roads that zigzag over Transmitter Hill. This wasn't an easy task, as many of these roads don't appear on any topographic or state map. Many of the residences on this hill were deeded to families decades ago, and many contain old trailers and homes that are now vacant and in ruins. It is not unusual to spend hours driving roads and eventually reach the end of one to find a dilapidated old residence that is now vacant. When you are that far off the grid or map, it is sometimes a very strange feeling to exit your vehicle and see that nobody has lived there for decades. Each time this happened, I always had the feeling that someone was watching as I walked the property and searched for Bigfoot clues.

It was a cold morning when I started the day knowing that I would be driving and meandering the roads of Transmitter Hill in search of residences. I started by driving the perimeter road that borders the Trinity on the west side of the river and driving up Transmitter Hill Road looking for the first dirt road that appeared to be drivable in my Jeep. I soon came across a nice road that appeared to be passable and in an optimum location. I drove this for approximately two miles and found that a large tree had fallen and blocked the road, a typical issue in these mountains. I made a U-turn and made my way back to the main road. Enroute to the primary road I saw a tribal forestry truck coming in the other direction so I pulled to the side of the road. I explained why I was up in these parts and that a tree was blocking the road just ahead. I asked if he could direct me to any residences in the area, and he said that he could. The tribal forestry employee stated that just above us ran another road and there was one residence that was actually lower than us but I had to go up higher to get the road. It was a complicated instruction and I took notes.

I found the main dirt road that was explained to me and made my way

Christy Brown and James Marshall at their residence.

on a very small, narrow dirt road that started to point towards the river. I soon came upon a very old trailer that appeared to be burned out and empty. As is typical in these mountains, there are several forks in a dirt road and you must take each one in order not to miss any opportunity. This was a very foggy and misty day, which just added to the odd feeling that I continuously had as I drove the roads. As I continued down toward the river, the road became extremely rough, the forest was dark and the odd feelings continued. After approximately 15 minutes of driving I came across a sign that stated "Private Property, Stay Out." This is always the point in any investigator's day where he decides to either go home or plunge ahead. My vote was to keep going.

Another mile down the road I came across a small house in a small meadow that had one vehicle parked in the front yard. The vehicle wasn't too old and looked operable, except for the broken windshield. There were kids' toys in the front yard and some garbage thrown throughout the yard, typical for this area.

I parked my car, took my clipboard and approached the front door. After several knocks and many seconds patiently waiting, James Marshall answered the door. He appeared young and it looked like I just woke him up. It was approximately 10:00 a.m.

I introduced myself to James and stated that I was investigating Bigfoot issues in the area and wanted to know if he could answer a few questions. James was very polite and stated that he would be happy to, but also asked if I could wait while he got his girlfriend to also meet me. James went into his house and quickly returned with Christy Brown.

James introduced Christy and explained that he had inherited the house from his late grandfather. He told me that the residence had been vacant for close to fifteen years until he and Christy occupied the house earlier that year (2006). Both stated that they have had a series of strange occurrences around thc house since they moved in. They explained that they have a motion sensor light on the tree near their vehicle. The light is on whenever they come home indicating that something is moving around the front yard. James said that they have two dogs but the dogs never go far from the house, and they appear afraid to go anywhere outside the confines of the cement sidewalks.

James and Christy both stated that she had been having medical problems the previous several months causing her to menstruate, sometimes continuously. This had been ongoing since early August and they were having her seen by a doctor. They stated that the first very odd issue occurred in mid August when they filled a small plastic pool that they had sitting in the front yard with water. The pool was approximately six feet in diameter and about 18 inches deep. They stated that the pool was kept full when they left for the day, any day. One day they left in the morning and returned later and found the pool overturned. They stated that they can tell when someone has driven down their dirt driveway, and they could see that nobody had visited. They maintained that there is no way their dogs could turn the pool over, as it's too heavy. It should be noted here that I saw the dogs and each probably weighs no more than five or six pounds.

In early September they were driving home from Hoopa at approximately 9:00 p.m. They stated that they drove up to their house and again noted that their front motion light was on. They exited their car and started to take their baby from the backseat when they heard the loudest scream they have ever heard in their lives. They stated that the scream was not like any animal that they thought lives in their hills. James told me that he has lived in the Hoopa area his entire life and knows all animals that live in the habitat, and he is certain that the scream did not come from any of the animals that everyone thinks lives in their woods.

They said the scream lasted 20 seconds. They guessed that the creature making the scream was 300 feet from their house. They claimed that they got their baby, entered their house and locked the door. They haven't heard the scream since.

In early October they were asleep when they heard something chasing their dog around the house. They say that they heard their dog bark just briefly as it was running around the residence. They took me to the rear of the house and you could see that the rear portion of the living area extends in almost a stilt-like fashion and there is room to run under the house if a human bends over. Both claim that the creature chasing their dogs was very large because you could hear very large and heavy foot stomps as the chase continued. Both say that they were too scared to look outside because neither has a firearm. After several minutes the chase stopped and they thought everything would return to normal. Christy explained that they now heard a slight squeaking sound and then saw their rear door bend slightly inward as though someone or something with great strength was pushing on the door. This door is at least eight feet suspended in the air with no stairs or ladders leading to it. James explained that they never got around to building the stairs, so the door is never opened. They say that the pushing on the door continued for 15–20 seconds and then stopped. They state that they heard loud and heavy footsteps leaving their residence once the pushing on the door stopped. Both Christy and and James said that the oddest part of the chase and the pushing on the door was that there was a distinct smell in the air, a smell that was worse than a skunk, a complete stink odor.

On Halloween day Christy and James said that they went to Hoopa so their child could trick or treat. At about 10:00 p.m. they returned to their residence in one of their cars and found that the car they left in front of their residence had a broken windshield. They said that they had purchased small pumpkins and had put them around their house for the holidays. They found that one of the pumpkins had been thrown with such force that it broke the front windshield on their car. Both Christy and James say that

Christy's and James' car with a broken windshield.

they know that nobody had come to their house that day, and it is baffling to think of what threw the pumpkins. It would have taken a very powerful throw to get a pumpkin to break the windshield.

Christy and James explained that they live on their road because it was inherited and because of its remote nature. I can definitely confirm that this residence is very remote. It would take a big effort to find the house, and you wouldn't accidentally stumble onto it if you were lost.

I asked Christy and James what they thought might be causing the problems at their house. After a few seconds both Christy and James said that it must be Bigfoot. They state that there is no other animal anywhere that they know about that is tall enough to push on their back door, and there is nothing in the backyard that something or someone could step onto in order to reach the back door. Whatever was pushing on the door had to be at least seven feet tall with an additional reach of several feet.

Location

After questioning and listening to Christy and James, I asked them if I could walk through their property and meander down to the river. They both stated that I could and they recommended that I stay on the dirt trail so I wouldn't get lost. I thanked them for their time, took their photo and put my notes in my car. I grabbed a water, my firearm (for bears and mountain lions) and started the downhill walk.

It was a good 20-minute walk until I could see the Trinity River. The path was very wide, well worn with no cover. I passed rows and rows of berry bushes, huge bushes. I also passed at least five huge bear scat piles

Den above James' and Christy's residence.

that were on the trail. During this walk I continued to feel uncomfortable and was constantly looking over my shoulder and around trees. I truly can't explain why I felt odd, it was just different. The cloud cover, fog and mist probably added to the scenario.

I finally reached a point where I could see the river and identify the specific location. I was at on the western bank. I was just downstream from the area where Debbie Carpenter's husband, Winkle, had seen footprints and heard a scream. The river in this location had a small beach and a huge area where you could walk over rocks and have easy access to a large portion of the Trinity. It was at the far northern end of the area where Debbie Carpenter lived and her sighting location of Bigfoot. This location was another piece to the puzzle. I could now draw a straight line on a map of sightings from Transmitter Hill to Bald Hills to Mill Creek. This location is approximately 1,700 feet down the hill from the Warrior Sanchez residence.

The walk out was very different than the walk in. It took me almost 45 minutes to reach my vehicle. I downed my water, rested and started the uphill drive through the forest. I was approximately 10 minutes from the residence when I noticed a large grouping of bushes down a small hillside. I stopped my vehicle and walked downhill to the location. These bushes closely resembled the den/nest I observed on Warrior Sanchez' property. There were very large trees that were pulled to a center point and then held in position by another tree. All of these trees made a natural area of cover. The area under the cover was packed down as though something had been lying there. There were no tracks that were visible on the ground because of the large amount leaves and brush in the area. This area is visible from the road and it has a full view of a large amount of the hillside and lower residence. You could almost stay in the den and have a full 180-degree view of anything that moved on the lower hillside. A vast amount of all of the brush and trees that were used in this shelter were still living, almost thriving.

I made my way back to my vehicle and continued my drive out of the area. It didn't quite strike me as wet when I drove in, but I did notice how damp the entire area around the hillside of James' and Christy's residence was. There were several springs that supported a lot of lush vegetation.

Scott Woodland

Masonry Contractor
Volunteer Tracker for local Sheriff's Office
Oroville

Date of Incident: 2006 (non-Hoopa related)

Every Bigfoot book must have some story about a Bigfoot cast and print. Although this specific story doesn't come out of Hoopa, it does emanate from Northern California. I am including this story in the book, as it is the only one in several years that I have researched that had possibilities as a real series of Bigfoot prints. This is to emphasize how rare it is to find a series of prints that can possibly be attributed to Bigfoot in this region of California. The weather and soil conditions make the finding and the preservation of the track difficult. This print also falls into the same time frame I was in Hoopa. I had found one track while I was in Hoopa and that was at the end of Lower Mill Creek Road and down a small path adjacent to the creek. The print was submerged in a small pond adjacent to the creek and it could not be cast. Casting a print is not an easy task, and it takes time and patience. The more that you rush the casting process, the worse the outcome.

Scott Woodland at the site where the prints were taken.

Over the last several years I have heard of several people who had found Bigfoot tracks in California. As I looked into each site and the issues surrounding the find, each time the tracks or the people or the location lost credibility and I chose not to list them here. I've heard other Bigfoot related professionals state that finding a track in the wild is as rare as making a Bigfoot sighting, and I'd agree. The forest floor in most of California is not conducive to retaining tracks for long periods of time. The floor is either riddled with leaves, pine needles or rocks. In the most extreme locations of California the floor is granite, granite everywhere. Much of the Sierra Nevada Mountains above 5,000 feet of elevation is either granite, lava rocks or so rocky that it would make prints impossible to find. The composition of the ground is a primary factor in where I travel and where I look for evidence. Granite is not likely to hold any evidence once the wind blows. Any hair fibers left on granite are gone at the first storm or wind. A very rocky path is unlikely to hold a print, but may hold other evidence. A dusty and dirty environment will hold prints for a few days, but then they'll lose finite texture. The best location for finding prints is along water sources, creeks, lakes, streams, rivers, ponds, springs and freshly carved roads. The sand and mud in these environments yield interesting prints from all types of animals. Since we are always looking for indicators (bears, etc.) finding their prints is very valuable.

In February 2007 I was contacted by a friend in Sacramento who told me about a news broadcast he saw claiming that two individuals had found

Bigfoot tracks in the hills outside Oroville, California. He gave me the dates, names and approximate location of the findings, and said that I should contact one of the witnesses who claimed to be a professional tracker.

In mid March 2007 I was able to contact the tracker listed in the story, Scott Woodland. Scott was very polite and helpful, and agreed to meet me at the site of the tracks. Scott said that he would meet me in Woodleaf, California and then lead me into the site where he was shown the tracks. He explained that a friend of his was at a site where they were conducting a logging operation. The logging was nearing completion and they were at the region after a long weekend. The crew found approximately 20 tracks leading down one of the logging roads. He explained that some of the tracks were fairly clear in the newly cut dusty road. He stated that his friend was able to cast one of the prints and he agreed to bring the cast to show me. He also stated that he took almost 200 photos of the prints and would bring a copy for me.

In early April I met Scott in Woodleaf, a very small town between Oroville and La Porte. There are no stores, gas stations or other services in this small town. There are several logging operations in the area and there is a continuous stream of large trucks that make their way up and down the highway.

Scott led the way up a northern dirt road from Woodleaf. We traveled approximately two miles and then turned onto United States Forest Service road number 19N07 and took this for another two miles. The first left turn with a gate was our last turn. We drove down an extremely narrow road that had very dense forests on both sides. The forest in this area was so dense that you couldn't see 10 feet in on either side. We traveled down this road approximately one mile until we reached a small clearing where Scott exited his vehicle. We were near Mt. Hope, the largest mountain in the area.

Scott now stated that we had to walk up the hill and down the logging road to a ridge where the road leveled off. It was a walk of about 500 feet. Once at the top of the ridge, Scott indicated that the tracks came out of a group of trees to the north, followed the road down the ridgeline and then disappeared among some downed smaller trees that were in a pile. He stated that he saw approximately 20 footprints. He claimed that all of the prints were approximately the same size, but he felt that one foot was slightly larger than the other, and that one foot had characteristics different than the other.

Scott told me that he was a professional tracker who had taken several classes on how to track and what to look for while tracking. He said that he works for a local sheriff's department in their search and rescue department. He stated that he has read a lot about Bigfoot, and has even communicated with a few professionals and has questioned them about the prints that were found.

Scott explained that when he looked at the series of tracks from a distance, and looking at the direction that they traveled, you could see that the left and right feet were offset more than most Bigfoot tracks. Scott emphasized that the area these prints were located was an area that nobody would walk in bare feet. There were rocky conditions with some garbage and metal that would make it very hazardous to walk in feet that were not covered.

The last item that Scott and I discussed was the cast itself. The cast appeared to be made in haste by someone who had no experience in casts. There were thousands of small bubbles throughout all layers and angles that made observing finite details impossible. There was one issue about the physical structure of the foot that was very odd. Just behind the pad of the big toe was another joint, a large joint that I'd never seen before. After the second joint was the beginning portion of the ball of the foot. This second joint could be seen on each of the first two toes. As I looked further at the cast I could see some type of line across the back portion of the ball that ran in a semi circle radius, which is another item I had never seen on a Bigfoot print.

I should state here that I do not claim to be an expert on Bigfoot casts, the foot of the Bigfoot or the structure at the bottom of the foot. I have seen, examined and researched many Bigfoot casts, and this cast had several aspects that were quite different than others. The condition of the cast made the identification of ridges, small identifiers and other aspects of the structure impossible to view.

Scott gave me his opinion on the differences between the feet, their size and description. He felt that the creature had one foot where its toes were scrunched up while it was walking which could have been caused by some type of injury in the calf, leg or foot. He explained that the actual size of the foot changed as the tracks continued down the road. He stated that he measured the variance between 16½–17½ inches. He stated that the right foot appeared relaxed and the left foot was tensed up.

Scott stated that when he was first on this site there were many signs of game, bears, deer and smaller creatures. He said that he found lots of bear tracks and lots of scat in the road.

There were several indicators about these prints that made them very appealing.

The Location

The elevation of this sighting was just over 2,600 feet, well within my hypothesis that Bigfoot lives and roams regularly in this elevation range in this region of the Sierra.

The forests and cover in this area are extremely thick. There were areas I passed while I was driving into the site where I couldn't see 10 feet into the woods. There was old growth in the area as well as good covering on the floor. There was ample cover for deer, bears and Bigfoot.

There were no homes anywhere within three miles. As I was driving into the site, and when leaving it, I didn't see any residents anywhere in the area or near the dirt road. This is one indicator that Bigfoot may have made this a regular stop on its travels and it felt comfortable being in the area.

There was a good water source on site. The water management authority in this area uses small canals to move water around on the mountains. There is a small canal, six feet wide and three feet deep that was running high when I was there. This canal ran the perimeter of this site.

The tracks were found on the dirt road that ran down the middle of the ridgeline of the property. I believe that Bigfoot likes to travel on ridges as this gives them an advantage for ambush killing and escape from people. This location of the prints fits with my hypothesis on travel.

The location had indications of many bears in the area. This is again one of my hypotheses associated with Bigfoot in Northern California. The more bears that live in the area, the higher the chance that the area will sustain Bigfoot. Bigfoot roam vast areas on a regular basis, but the area where they primarily reside appears to be a location that sustains a high population of bears. There are too many indicators to ignore this possibility.

There have been many Bigfoot sightings in this region of California in the last 60 years and the locating of Bigfoot prints has been very sporadic in the last 20 years. Once the road building boom of the 1950s and 1960s subsided, it was much more difficult to find Bigfoot tracks. The cutting of a new dirt road makes an ideal location to spot new tracks of any creature. Once road building subsided in the California mountains, it became much more difficult to spot Bigfoot prints.

The Print

I have never seen a print with an extra digit between the pad of the big toe and the ball of the foot.

I have never seen a print with what appeared to be a line between the pad and arch that formed a semi circle type pattern across the foot.

There was a large indentation in the arch of the foot on the cast, not a normal characteristic in the prints I have examined.

The cast was poorly made. There were so many bubbles in the cast that detail in the foot was impossible to identify. It almost appeared as though the casting material was put into a blender and then poured onto the ground.

Conclusion

Scott made several great observations about the conditions and environment of the site. I think that his professional ability as a tracker brought significant credibility to this print. Scott did not cast the print, but did show me the location where the cast was taken, and the location of other prints still could be faintly seen.

If this track was a fake, then the person who pulled the prank had extraordinary knowledge of Bigfoot habitat and took significant time in finding the optimum location (ridgeline, habitat, water, bear, game, elevation) to place the fake.

If I do locate another print that has a double digit behind the big toe, I'll certainly remember Scott's cast.

Chapter 5
Sightings Involving Bigfoot

For Sighting #1 refer to Sighting #6, page 174

Josephine Peters

Retired Herbal Specialist
Hoopa

Sightings #2: 1963
#32: January 2006

Ed Masten told me that he had an aunt (Josephine Peters) he wanted me to meet who was a witness to multiple Bigfoot encounters. He said that she lives adjacent to Supply Creek and next to the youth recreation center.

I met with Josephine at her residence. She lives in a small house on a very large lot. There is approximately 300 feet between her front door and the street, Loop Road. The lot has nothing on it except the residence and trees immediately adjacent to the house. The front of the house is an open gravel field with Supply Creek running less than 200 yards from her home. The creek starts in the hills that are between Hoopa and the Pacific Ocean.

Josephine told me that she was a Kurok and Shasta tribe member, but had lived in the Hoopa Valley for 57 years. She was born in Somes Bar, California but has found the lifestyle and climate in Hoopa to be more conducive to her lifestyle. Josephine told me that she's had two Bigfoot experiences in her life, but she has always heard that Bigfoot lived in their area and frequented the Hoopa Valley. The first encounter was when she was much younger and near the Klamath River. The most recent event occurred early 2006 in her front yard.

Josephine Peters sitting at her kitchen table.

Josephine first heard about Bigfoot from her grandmother. She distinctly remembers her grandmother giving directions to her sons about camping. The sons were going fishing and were contemplating where they would camp near the forks of the Salmon River. She heard their discussions and stated loudly, "You cannot stay at Crapo Meadows at the Forks of the Salmon, that is the Big People's area." Josephine stated that her grandmother was adamant, upset and vocal. Her sons took the advice and stayed out of the area. She said that the story has stayed with her all these years. The Salmon is one of the most beautiful rivers in Northern California and is used extensively for whitewater kayakers. The region around the forks has few tourists because it takes so long to get to it from any metropolitan area (seven hour drive from San Francisco). The Forks are surrounded by wilderness area, thus making traveling the landscape slow and difficult. The region is lush and appears to be ideal habitat for Bigfoot.

The first encounter between Bigfoot and Josephine occurred 10 miles east of Weitchpec off Highway 96 at Slate Creek. The creek is a year-

Salmon RIver.

round stream that is easily identified by its rusty colored water that is caused by a local iron deposit. There is a road that follows the creek north from the highway but it leaves the streambed area immediately and takes you hundreds of yards from the water in less than a mile. The only way to travel up Slate Creek is to enter the road from the Highway 96 overpass and work the streambeds as you walk up. After a two-hour dusty and rough drive on this road, you will end up in the Bluff Creek Region.

Josephine stated that she was making her way (walking) upstream with a few friends looking for herbs. Josephine is known in the tribe as a specialist with herbs. She stated she was probably the furthest person upstream when she thought she heard something playing in the water not far from her. She looked upstream and saw a huge creature several hundred feet up from her, in the middle of the creek. She said that the creature had the standard Bigfoot appearance, hair covering the entire body, standing on two feet, face similar to a human and not a bear, and using its hands like a human. She stated that the creature was huge, hundreds of

Josephine Peters' front gate where she spoke to Bigfoot.

pounds, and probably over seven feet tall. She quickly and cautiously made her way back down stream, notified her friends and they quickly left the area. She said that she thinks she is the only one to have seen the creature.

Josephine explained that her next Bigfoot observation would be decades after the Slate Creek incident. She said that it occurred in late January or early February 2006. It was approximately 10:00 p.m., and Josephine heard her dog barking in the front yard. The dog continued to bark as it was making its way at full sprint around the house, into the backyard and then under the house. She recalls this was very odd behavior for the dog because it had never shown any cowardice towards anything in the past. Josephine opened her front door and walked out to the fence that surrounds her yard. It was then that she saw a huge Bigfoot, its silhouette standing behind her truck that was parked near her house. She stated that she told the creature to come closer, which it did.

I know this seems like incredible behavior for an elderly person to exhibit, but Josephine struck me as a very levelheaded woman who was not afraid of anything — seriously. She had told me that she had led a full life and she had never heard anything to indicate that Bigfoot would hurt anyone, and she never knew if she would ever see the creature again. She wanted to make as much out of the encounter as possible.

Josephine stated that the creature walked to the edge of the fence and stood staring at her. She said that the Bigfoot looked like the Bigfoot seen on the Patterson film, except this creature did not have hair on its face. It was huge, calm and made no noise.

She did make the point that the creature had a horrible odor. The closest description she can give for the smell is that it was similar to an old, wet raccoon. She said that it had greenish colored eyes that were very captivating. She attempted to elicit conversation, but the creature said nothing. She said that she did not have food with her, but left and went to get some from the kitchen. She came back with a loaf of bread, but found the creature had moved out of sight. She left the bread where the creature was last standing and went back into the house.

An hour later she found that the creature had taken the bread. Josephine said it is her belief that Bigfoot will not eat heated food and prefers natural, cold food. She said that she has left food out in the past and the creature never takes any type of heated food, or moist dog and cat food.

Josephine said that a few days after her last encounter, she was talking with some local boys that were standing on the bridge over Supply Creek, which she can see from her house. The boys started to tell her how they had seen a Bigfoot two days earlier in the late afternoon in the creek below the bridge. She told them that it had come to her house that very day.

As I was leaving Josephine's residence I was given ritual Native American herb to give me good luck. Her friend told me that this would guarantee my safe journey and it would also ensure that I would see Bigfoot soon. Josephine's parting words were that she has never heard a story from any of the three generations of Hoopa family that Bigfoot has ever hurt anyone. She told me that she isn't afraid and never will be afraid of Bigfoot. I believed her!

Forensic Sketch

Josephine was very helpful and supportive of doing a forensic sketch of the creature that came into her front yard. She arrived with her niece at the hotel room and took a seat across from Harvey Pratt, forensic artist. She started the conversation by talking about both of her sightings and explaining her life in Hoopa. Josephine also talked about the practices of the Hoopa elders.

Josephine explained her sighting to Harvey as lasting approximately five minutes at a distance of 20 feet. She stated that the creature was approximately six feet, six inches tall, with broad shoulders and it had a strong bad odor. She said that it was basically brown in color with streaks of grey, and green eyes that glowed. She said that the creature stood in a pigeon pointed stance and had hair on the back of its hands. Josephine said that they have a motion sensor in their front yard and for some reason it came on while she was in the front yard with the creature, not when the creature originally walked up. When the light came on, the creature took several steps backwards. She stated that she tried to talk to it and it just grunted when she stopped talking. Josephine stated that she has left out dried cat food for the creature and it is taken after she leaves it. She's not positive if it's Bigfoot or another creature that eats the food.

Harvey and I did ask Josephine about the history of the Hoopa people and the relationship between the people and Bigfoot. We specifically asked why Bigfoot has so many encounters with people in Hoopa. Josephine replied that when she was much younger her grandparents told her that their parents did things when they were younger that they were not proud of. She stated that in the 1800s life was very difficult for the Native American, and survival was at the core of their existence. She said that when a Native American woman had a child that was severely disabled or sick, rather than watch the child die, they would take the child up into the mountains and leave it. She stated that this was something that wasn't talked about at length, but people had always wondered what happened to these children. If you believe that Bigfoot is the man in the woods that takes care of the mountains and the forests (as Hoopa tribal members feel today), then was it Bigfoot that took care of these children? Bigfoot still has curiosity in children, as is noted in numerous sightings. The real ques-

Harvey Pratt's sketch of the creature that visited Josephine.

tion that comes to mind is, did Bigfoot breed with these children and is that the core reason that Bigfoot has a face that is close to human?

Many tribal members from Yurok and Hoopa have told me that tribal members regularly went to the mountains to escape wars and fights with different enemies. Josephine was the first person who told me about leaving children in the woods.

After Harvey finished the sketch, Josephine looked at it and was amazed at how close the sketch matched the creature in her front yard. Josephine signed an affidavit.

The sketch of Josephine's creature is important for a variety of reasons. One fascinating aspect of the sighting is that there are few witnesses on record who have had an encounter with Bigfoot for the same duration as Josephine's. Most sightings are seconds and not minutes. Another aspect of the sighting to remember is the state of mind that Josephine had when she went into the encounter. She knew she might never see the creature again and wanted to make all she could of the event. With a mindset of seeing everything possible and making the most of the encounter, a witness has the opportunity to look for detail where most people would not. Harvey made a comment after Josephine left that he felt she was incredibly believable and great with the facial details. We focused on the face of the creature with Josephine because this was an exceptional opportunity with a very brave lady.

Marion "Inker" McCovey

Recreation Director
Hoopa

Sightings #3: 1964
#16: November 2003
#23: November 2004

When I initially met the Tribal Police in Hoopa, they referred me to Inker. The officers stated that "Ink" is a believer in Bigfoot and had seen the creature in the past. They told me I could contact him at the recreation center, and advised me that he was a central figure in the community, and a credible person.

I drove to the recreation complex near the center of Hoopa. It is a modern, beautiful facility that includes a pool, play yard, daycare facility and indoor basketball court with workout equipment. I found Inker playing hoops with kids inside the facility. He is a tall, large man, dark complexion, pony tail and obviously from a Native American background. He greeted me with a big smile and offered his assistance. It was obvious to anyone within a hundred miles that I was not from the tribal area — too white, too many questions, and lost. I explained to Inker I was in Hoopa investigating

Bigfoot and asked if I could talk to him. He told the kids that the facility was closing soon and that he'd be in his office talking to a guest.

We sat in a modern office (with an Internet link) overlooking the basketball court. Inker immediately started the conversation by stating that his history with Bigfoot began in his childhood. He said that his family owned 165 acres at the foot of a nearby mountain and that his entire family resides on land parcels in that area. He said that when he was eight years old, in 1964, he went to the bathroom at his house and was sitting on the toilet. He said it was a warm summer evening and the bathroom window was open, as it was normally. The window was approximately six feet up off the ground so people could not look inside. He stated that he was sitting on the toilet when he saw a huge, hairy arm come through the window and reach toward where he was sitting. Inker stated that the arm was much larger than a man's in girth and length, and much, more hairy than any human arm. The arm appeared to be reaching toward him and he started to scream. His father came running into the bathroom and the creature pulled its arm away.

Inker's dad grabbed his rifle and a flashlight and ran out the front door searching for the creature, but could not find it. He said that he can distinctly remember that the creature had grey/black hair on its forearm and the hair was almost four inches long.

After explaining the bathroom story to me, Inker paused a few seconds and said that all local tribe members believe that Bigfoot is here to take care of the forests. They are not here to do any harm or destruction. He said that they should be respected and left alone and in peace.

In November 2003 Inker said that he and his girlfriend at the time, Mary McClelland, were driving from his residence to the local casino at approximately midnight. They were taking their normal route down Mill Creek Road and turning left onto Highway 96. He stated that he was driving and his girlfriend was in the right front seat. As they made the turn, they immediately saw a Bigfoot on the Trinity River side of the road. It was at least seven feet tall, standing on two legs, and looking in their direction. He stated that the high beams of the car illuminated the animal and they could see its orange-red eyes. Once the headlights hit the animal, it turned and took two huge strides across the highway towards Big Hill Mountain and then leaped over a 10-foot-tall cyclone fence and disappeared.

Inker knew that he and Mary had just seen Bigfoot. He said that the creature had hair over its entire body, walked on two legs and obviously did not enjoy the light coming from his headlights. The creature had a huge body with a massive build.

Inker stated that many members of the tribe and his family have seen Bigfoot. He said that the property that his family owns is ideally situated at the bottom of Big Hill Mountain and a small drainage basin of Mill Creek. He explained that approximately six years before there was a huge fire behind Big Hill that caused significant forest destruction (Megram fire). Inker stated that since that fire, the creature has made its presence known on a more regular schedule, and it appears that its habitat has moved closer.

Harvey's sketch of the arm that reached through Inker's bathroom window.

McCovey stated that two years past (November 2004) his mother had a bad stroke. He was depressed and decided to hike into Big Mountain with the primary intent on collecting tan oak mushrooms, keeping his senses alert for any evidence of the creature. He said that he drove as far as possible up into the Mill Creek basin and then walked a small ridgeline. He had collected two large baskets of mushrooms and had not seen anything unusual. He was walking on a small ridgeline near an adjacent small mountain when he felt something was staring at him. Inker said that he felt very uneasy and decided to sit down, look and listen, something tribal elders had taught him to do when in the woods (a common behavior that all Natives practice). He stated that he first looked at the opposite mountainside approximately 100 yards away. He thought he saw a small black tree stump move. Inker said that he didn't have binoculars, but he did have his rifle and scope. He looked at the object through his scope and saw a large fir tree behind the object also move. He cleared his eyes and re-focused on the objects. He saw the small stump-like object continue to move and could now see that it was a small, adolescent Bigfoot with a large adult Bigfoot directly behind it leaning against a tree.

Inker's sketch of an adolescent and adult Bigfoot near tree.

As he was looking at the creatures, the adult swung its leg in a semicircle motion and concealed itself almost entirely behind a large fir tree. He said that the wind was blowing from the creatures towards him and that he could distinctly smell a foul odor, a stink. The odor smelled like strong ammonia, or a very strong pissy smell. Inker said that he continued to watch the creatures for a few minutes and felt as though the adult was telling the juvenile to continue to stay in a ball and not move or get hurt — at least that was Inker's sense of the situation. A thought to remember at this point is that McCovey was looking at the creatures through a riflescope mounted on his rifle. It would be easy to understand the perception of the creature believing that a shot might be taken at them in a mistaken belief that they were game.

After several minutes of watching the creatures, he started to feel uncomfortable. He said that he dumped out the contents of one of his buckets and put his accumulated mushrooms in a large pile where the creatures could see them. He then left the area and did not come back. Inker said that many people in the community feel that Bigfoot feed on the mushrooms on the hillsides, and that is one of the reasons they have stayed close over the years. He said that this is also why many people have seen Bigfoot digging in lawns and grounds, looking for mushrooms. Inker said that he wanted to apologize in some way to the creatures for disturbing them, thus he left the mushrooms as a goodwill gesture.

I explained to Inker that I would like to enter an area of Bigfoot habitat that is rarely explored, yet is frequented regularly by the Bigfoot creature. Inker had a broad smile, hesitated slightly, and then stated, "I know the exact spot." He told me that on the backside of Big Mountain there is a ridgeline that travels all the way to the back of Bluff Creek Resort. This entire ridge parallels Highway 96 to Weitchpec.

Inker and his son Romeo at their residence.

He told me that he knews I was aware of the sightings along Highway 96 from Hoopa to Weitchpec, of Bigfoot coming off the hillside and down towards the Trinity River. Many people have made observations of the creature crossing Highway 96 in this area. Inker said that a few years ago he was hunting with a friend in the Shelton Butte area and had seen signs of Bigfoot activity. He said that nature trails had trees and limbs broken high up off the ground and there appeared to be a Bigfoot den that he had found. He explained that the area of Shelton Butte is a portion of the reservation that elders in the tribe have told others to stay away from. It is known as Bigfoot's area. Inker said the area includes part of the mountainside that is behind the Bluff Creek Resort, generally on the south side of the Klamath River. He said that the area is very steep, desolate and the perfect area for Bigfoot to travel to the Klamath River (Phil's Bluff Creek Resort area) and the Trinity River areas just outside Hoopa.

An interesting note: I had interviewed Inker after I had met Phil Smith at his Bluff Creek Resort. Phil had told me of seeing Bigfoot up the mountainside behind his resort. Phil's sighting matches the area that is described by Inker. Inker and Phil have never spoken, and don't know each other.

Inker finds it a little unusual that Bigfoot spends a lot of time near the Trinity and Klamath Rivers. He said that there is an old Indian legend that there is a serpent that lives in the rivers called Kamoss. He stated that Kamoss is a snake-like creature that has a head the size of a horse and presumably eats fish. Inker said that there is another Indian legend that tells kids not to go into the river until you can see its bottom, because elders do not want Kamoss to eat you. He said that there are a lot of people in the community who have seen Kamoss and know it is a real creature. He said that the serpent makes a very high-pitched motor sound at night, which can be heard over long distances. Inker said that one of the people who has seen Kamoss is Officer Chance Carpenter who saw the creature near Big Rock. He told me that it is a little unusual that Bigfoot and Kamoss live in the same area and are so elusive.

McCovey stated that he would gladly take me to his house and introduce his family to me. He said that there were many members of his extended family who had witnessed Bigfoot in their neighborhood. We left the recreation center and drove to Inker's house.

Forensic Sketch

Inker has always been a person that has made himself available to assist in this research. From the very beginning, he has stated that he wants to get all the information he can about Bigfoot and fully understand the creature and its habitat. He has also been a person that has kept me clear of trouble, both people and tribal. He truly seems to understand the landscape and the tribal politics.

The first sketch that Harvey developed with Inker was the arm that came in the window when he was a child. Inker stated that even though it occurred many years ago, the incident is fresh in his mind. Inker said that when the creature stuck its arm in, it reached to the point that the window frame stopped it from going in further and its bicep took up the entire window frame. He specifically stated that the bicep had the diameter of a basketball. Furthermore, the arm did not have hair on its inner section and its fingers were more ape-like than human. Inker said that the skin on the arm looked like dark leather. During the interview, he said that it was his feeling that the creature had somehow seen him enter the bathroom and it was specifically trying to grab him. He said that the sketch is a very good representation of the arm. He stated that he is 100 percent positive that it was not a human arm that reached for him.

After Harvey completed the arm sketch, Inker said that he would try to give the best description of what he saw on the mountainside. During the interview with Harvey, Inker stated that he felt the big Bigfoot was close to seven feet tall, but couldn't determine the height of the creature that was crouching. He did say that he felt that the two Bigfoot were communicating with each other, with the larger, or mature, Bigfoot telling the smaller one to stay in a ball and not to move. Inker told us that a few weeks after making this sighting he went to the mountain location where the Bigfoot had been standing. He stated that he went to the specific tree where the Bigfoot were standing and saw that the large Bigfoot was partially concealed in a burned out hollow of the tree it was standing behind. Inker was pleased with the drawing of the two Bigfoot and stated that he really couldn't add more detail because he wasn't close enough.

In the thousands of sighting reports I have read, numerous witnesses stated that they saw a large black stump or ball on a hillside, they looked at it momentarily and after a few seconds it started to move. It appears that Bigfoot has developed a life maintenance strategy of concealing itself in the landscape by staying motionless. This is a recurring theme in sighting reports, and Inker's observation reinforces this strategy as reality.

On September 7, 2006, I was sitting in a local Hoopa café eating breakfast when Inker McCovey walked up to my table. We exchanged greetings and Inker asked me what was happening in the community. I told him I was en route to contact Warrior Sanchez about a possible Bigfoot encounter. I told Inker that I wasn't positive about where Warrior worked, but that people at my hotel were helping me locate him. Inker stated that Warrior worked out of an office at the Hoopa newspaper and he would gladly take me there after breakfast.

Inker told me that one of his sons recently had a Bigfoot encounter with a group of other boys. Inker said that homes in his area had an ongoing problem with bears getting into their garbage, and that the problem had reached a crisis point and neighbors were going to shoot the bears when they found them.

One of Inker's sons had taken their .270-caliber rifle and, with a group of other boys, was going to walk up the road to Upper Mill Creek to find the bear. They were walking up the road when they saw something grey on the side of the road. Inker said that they were approximately 25 feet from the animal when it started to stand up. The creature did not turn toward them but away and back towards the upper hillside. He stated that they did not see its face, only its back. He said that the creature was over seven feet tall, had a grey upper back and was huge and walking on two feet. Inker said that the creature made a small grunt and started up the hillside.

Inker's son and his friends came directly back to Inker's house and notified their dad of what they had witnessed. He wasn't surprised.

I asked him if he was still willing to take me into the Hoopa property behind his house and show me the area that Bigfoot lives. He said that he was eager to do this and he was sure that he would gather further proof of Bigfoot's existence. I asked Inker why he was so sure that Bigfoot was living in this area. He said that the Shelton and Hopkins Butte area has been known by generations of Hoopa people as an area where Bigfoot lives. He said that the Hoopa people steer clear of this region in an effort to give Bigfoot his living area. He said that he has always been interested in Bigfoot and has walked through this area a few times in the past. Inker said that one time he was deep inside this region when he was walking on a game trail. He stated that it immediately struck him as a very peculiar trail because it was very wide, almost five feet and the canopy of the trail was over seven feet tall (an almost identical statement as to what Tony Hacking made about game trails). A normal canopy on a game trail isn't over five feet, based on the height of a deer, bear or sheep. Inker also said that he noticed several 4–6-inch-thick trees along the trail, which had been twisted and broken off approximately six feet off the ground. He said that the entire time he was on the trail he felt uncomfortable and also felt as though he was being watched.

I asked Inker about water and food sources that would be available to the Bigfoot in the Butte area. He said that there are a lot of deer and bears. It has long been the opinion of tribal elders that Bigfoot eats mushrooms that proliferate on the hillsides around Hoopa. Inker also said that the region has many springs that run year round, in addition to small creeks and the Klamath River that is on the backside of the Buttes. This region is on the back of the Bluff Creek Resort and falls in the area where Phil Smith (Bluff Creek Resort) had made his observations of Bigfoot.

Inker claimed that each Butte has many rock outcroppings that hold many caves. He believes that Bigfoot families live in these caves and then migrate at night down to the Trinity and Klamath Rivers. Inker said that he is positive that we will make observations that will show that Bigfoot resides in this area.

Romeo McCovey

Student, Hoopa High

Sighting #29: November 2005

Romeo was invited into the living room and introduced to me by Inker, who started the conversation by telling me that Romeo was his son and was 17 years old.

Romeo told me that he had heard Bigfoot stories his entire life from a variety of family, elders, friends, teachers and even his dad, and didn't believe any of them. He said that he never really thought much about it, and it was funny when people around him talked about it.

Romeo now told me his personal story. It was November of 2005 when he was driving down from Highway 96 and turning onto Mill Creek Road. He stated that he was alone and it was early in the evening. It was cold outside, but no snow on the ground. He stated that he had just turned onto Mill Creek Road when he saw something hunched over at the creek side of the street. He stated that he slowed to 20 mph and his car headlights were on the object. As the light illuminated the creature, it stood up on two legs, was approximately seven feet tall, with very wide shoulders and a human-type face and hair covering its entire body. He said that he knew he was looking at Bigfoot. Romeo raced home and immediately found his dad.

Inker said that when Romeo came in the door he was very shaken. He said it is rare to have Romeo that disturbed about anything; he just isn't that type of kid. Inker stated that hours afterwards Romeo was still so disturbed by the incident that he slept on the couch at the foot of Inker's bed, something he hadn't done since he was an infant. Inker plainly stated, "The kid saw Bigfoot."

Location of Sighting

The area of Romeo's sighting is significant for many reasons. The region of Mill Creek and Highway 96 has had an abundance of Bigfoot sightings, thus I coined it Bigfoot Alley. In this specific area there are many large berry bushes with Mill Creek harboring a large number of fish. Bigfoot has a long history of visiting this specific creek valley, going all the way back to when Inker was a child.

Inker and Romeo each signed an affidavit.

Forensic Sketch

Romeo met with Harvey Pratt and worked extensively on the sketch. It became evident early in the interview that Romeo definitely saw a Bigfoot but didn't see it in enough light to add significant detail. He did state that it did not appear to have a neck and it had wide shoulders. Romeo estimated its height at over eight feet and it had shaggy hair over its entire body. He said that the best he could determine, the creature acted as though it wasn't even observed and appeared unconcerned. Romeo explained that the creature appeared to have a very powerful body.

Harvey's sketch of Romeo McCovey sighting.

Harvey worked the best he could with the limited information and description given by Romeo. The finished product was witnessed by Romeo, who indicated that it was the best representation possible of what he saw.

Jesse Allen

U.S. Postmaster
Forks of Salmon

Nancy Allen

Sawyers Bar

Sighting #4: July/August 1967 or 1968

Many years ago I heard stories emanating from the region east of the Trinity Alps wilderness area regarding Bigfoot sightings. Many of the incidents were in isolated, desolate stretches of the wilderness that had little foot traffic, no vehicles and extremely remote campsites. The Forks of Salmon would remind you of a small city that architecture of the 20th century has forgotten. It is very isolated, remote and doesn't have a huge highway leading to it. It does have the beautiful Salmon River flowing on two sides. The Salmon is one of only two rivers in California that is truly wild; the other is the Smith River in far Northern California.

It is not an easy trip to get to the Forks of Salmon, which is one reason why it has a special charm and beauty. There are no fast-food restaurants, no 7-11s; in fact, there really are no commercialized establishments except for a small trailer that is the general store and a very small gas station. The only sign of industry in the entire area is a remnant of an old mine, and a small road crew warehouse for Siskiyou County Department of Public Works. There are several small campgrounds in the area and miles and miles of open river to fish and raft. I also passed a small elementary school that was in session and appeared to have 3–4 classrooms and probably 5–6 cars in the parking lot. They had a small, gorgeous field and play area for the kids, and all of this sat adjacent to the only roadway in or out of town.

After driving through the center of town, I looked for a location that appeared to house a governmental presence of any type. The lone building in the town that looked like 21st century construction was a tiny United States Post Office sitting alongside the river. It was maybe 300 square feet and had two parking spaces. There appeared to be someone working behind the counter and it looked to be open for business. I pulled

Jesse Allen at her job as postmaster in Forks of Salmon.

into one of the spaces, got out in a driving rainstorm and was warmly greeted by Ms. Jesse Allen. Jesse stated that she knew I wasn't from around there, and I was dressed too formally to be a forester. I handed her my card and stated that I was there looking into the Bigfoot phenomena. I asked her if she knew of people in the community who had seen the creature or who had stories of the creature. Jesse said that she did know people; in fact, Jesse said that she knew everyone in the area and had lived in the region for most of her 54 years. Jesse said that she was born and raised in Hoopa and was very familiar with Bigfoot and its associated stories and legends.

Jesse was very cautious during the initial minutes of our conversation. She asked a lot of questions that seemed to test my knowledge and sincerity in Bigfoot and the region. During her questioning she asked me to excuse her for doing her governmental duties as she was having a delivery and pickup around noon and needed to be prepared when it arrived. I told her that I respected her diligence and would be happy to work around her schedule. After approximately 20 minutes of Jesse working industriously, she apologized, came back to the window, looked me in the eye and started to talk about Bigfoot.

Jesse stated that she spent her entire youth living, playing and exploring the area around Forks of Salmon. She pointed across the river and around the corner and stated that she and her family lived in a house at that location the entire time she was growing up. Jesse said that she and her sister had attended the same elementary school that I had passed on the way into town, and that the school has changed little in the 40 years since she had attended.

Jesse explained that in July or August of 1967 or 1968, it was a ritual in Forks of Salmon for many of the townspeople and kids to go to the elementary school after dinner and play volleyball until after dark. She stated that the parking lot would be filled with cars of people who had to drive into town to play, and many others just walked. Jesse said there were times when the games went past dark and locals would park their cars around the courts to light it up so play could continue. This particular night in July or August, Jesse remembers her dad specifically telling her and her sister that they were to be home immediately after it became dark. She said that they had pushed the envelope on too many nights in the past and her dad was getting a little angry with them staying out too late.

She recalls that they got ready to leave their house at about 6:30 p.m., just after they had finished dinner. Jesse said that they walked down their long dirt driveway, reached the main roadway and made the left turn to walk towards the Cecilville Bridge over the Salmon River. Jesse stated that back then the bridge was much different than it is today. The bridge was made of all wood and creaked a lot when you drove or walked on it. She stated that you had to be careful when you crossed it in bare feet to not get nasty slivers. She can distinctly remember that she and Nancy (her sister) had one pair of flip-flops between the two of them, something that her parents would not have approved. She remembers walking carefully across the bridge to avoid any injury. She stated that she didn't like to wear shoes because she always took them off when she played volleyball.

The volleyball game continued until way too late into the night. She said that she knew her dad would be getting angry because she and Nancy were not home before it got dark. She and Nancy left the school at approximately 9:30 p.m. and were starting to work on their story to tell their dad about why they were so late. She said it was pretty easy to develop because it was a full moon and wasn't too dark when the sun had completely disappeared. It should be noted that the sun disappears pretty early in this region. The river has very steep canyons and the mountains are very steep, so the sun disappears early in the summer and winter.

Jesse said that she thinks that Nancy wore the flip-flops home. She said that she can remember thinking what a beautiful night it was with the moon full. As they were crossing the river the moon had lit up its riffles; it was gorgeous. It was summer and the river wasn't running full, but it was enough to make a gentle rumble. Jesse said that they finished crossing the bridge, walked the short stretch down the road to their driveway entrance and made the right turn into their property. They owned horses and sometimes they greeted them near the entrance, but not this night. Jesse said that not seeing horses was a little strange especially with the great nighttime visibility because of the full moon. Jesse said Nancy always walked looking at the ground, a habit she had developed as a small child, so Jesse always led the way by sometimes holding her hand. The girls continued to make their way into their property by walking up a slight grade in their driveway and then making their way downhill towards their house.

Jesse said that as she and Nancy were starting down the small knoll, she was looking towards a large oak tree that acted like a canopy over the driveway. Jesse was looking down the driveway towards the oak tree, which was being backlit by the full moon. She said that she saw the creature near the canopy of the tree. "It was clear to me what I was looking at, but my sister had her head down," Jesse stated. A huge figure was standing directly in the middle of her driveway under the oak tree. It was unbelievable.

Jesse looked at my business card and stated that the silhouette identically matched the figure of Bigfoot on my card. She told me that she had heard of Bigfoot as a kid, knew how big it could be and knew that she was looking at a Bigfoot at that moment. Jesse guessed that the creature was over seven feet tall, it had huge shoulders and was swaying slightly side to side on its two feet. She said that she could tell that the creature had either hair or fur over its entire body and that the color was dark, but that was about as far as the description would go. She said it was obvious to her that the Bigfoot was looking at them, but she couldn't see its face because it was dark. Jesse said that she now pulled on Nancy's arm and told her to look up. Jesse said that for some unknown reason, Nancy picked a rock up off the ground and threw it directly at the creature, a throw of about 30 feet. The rock hit the creature, bounced off and onto the ground. Nancy then yelled at the creature, "What are you doing here?" Jesse said that the creature made no sounds. There was no wind that night and they could not detect any odors.

Jesse said it was obvious that this huge creature wasn't going to move, and it was in a direct line for them to get into their house. She stated that they both were probably thinking that it wouldn't be too smart to continue to throw rocks, so they simultaneously turned and ran back to the main road. She said she ran the fastest that she has ever run in her life. Jesse said they both apparently knew where they were running to — Harvey Deala's residence that was on the opposite side of the Cecilville bridge. This was also the regional United States Forest Service Office and Harvey was the foreman. Jesse stated that it seemed like it took forever for them to get to the residence, but in reality it probably only took a few minutes.

Jesse said that she pounded on the Dealas' front door and Harvey answered. They told Harvey that they had just seen Bigfoot and they ran into the house. Jesse said that she can remember Harvey not believing their story, but feeling sympathetic enough that he would have his son Mike drive them home. Jesse said that Mike believed their story and was very reluctant to make the trip back to their house. He drove to their front gate, but wouldn't turn the car around and shine the lights down the driveway. He told the girls that he wasn't waiting around, and once they were out of the vehicle he immediately turned the car around and headed home.

Jesse remembers cautiously walking down the driveway and, seeing that the creature was no longer there, running at a flat-out sprint to their front door. She and Nancy ran into their house once their mom opened the door. They told their mom what they had seen and the trip they had made to the Dealas. Jesse then asked where dad was. Jesse said that their mom said that their dad had gone to bed over an hour before because he was exhausted.

Jesse said that she and Nancy rarely talk about the incident, but both

Nancy Allen just outside of Forks of Salmon.

agreed they had a close encounter with Bigfoot. She stated that in her years at Forks of Salmon she doesn't remember anyone else claiming they had a sighting in town, but did remember huge footprints found near Sawyers Bar over 10 years earlier. Jesse encouraged me to travel to Sawyers Bar and talk to her sister and compare their stories.

Jesse said that the trip down her driveway was never the same after that summer night of volleyball, rock throwing and Bigfoot.

Jesse gladly signed the affidavit and again encouraged me to contact Nancy for her side of the story. She called Nancy for me and we set a subsequent meeting.

Nancy and Jesse have similar demeanors and outlook on life. They are very polite people and quite easy to talk with. I initially asked Nancy to recount for me what had happened the night on the way home from volleyball. In short, Nancy's recollection of the events was extremely similar to Jesse's. Her memory of the event is still seared into her brain and she said it is something that she will never forget. The conversation with Nancy was on a road just outside Forks of Salmon in an absolutely beautiful environment. Nancy said that she and Jesse are in agreement that they believe they saw Bigfoot that night in front of their house. Nancy signed an affidavit.

Location of Sighting

Forks of Salmon is located almost due east from Hoopa and directly across the Trinity Alps Wilderness Area. The area west of the Forks is very desolate with no vehicular traffic allowed because of its wilderness status. The only way into the wilderness area other than walking is by horseback, a slow and expensive method of transportation. There have been many locals and Hoopa residents who believe that wilderness status for the Trinity area was granted specifically to give Bigfoot an area to live and thrive without being disturbed. This is an interesting theory because, even without the wilderness status, this region of the state is so remote that it was rarely visited prior to becoming a wilderness region. The only people normally found in these areas are loggers, hunters and some fishermen (and maybe Bigfoot researchers).

A Bigfoot standing in a driveway and greeting two girls coming home at night is not surprising. There have been many documented cases where Bigfoot makes its presence known to kids in what appears to be an attempt at interaction. It would appear from the documentation that Bigfoot knows that kids pose no threat, but attempts to act non-aggressive and invite the interaction utterly fails time and time again. The mere size of the creature would be enough to scare the most daring child. The reaction that the creature saw from Nancy and Jesse probably let the creature know that they were scared and it wasn't wanted on their property and that's the reason it never came back.

I am including the sightings at Forks of Salmon in this book because I believe that it all falls into the landscape of what Bigfoot would traverse in a normal week. It would make sense that a Bigfoot hunting for food and moving with the seasons would easily migrate and move from the Forks of Salmon, Orleans, Weitchpec, Hoopa, Bald Hills and may eventually make its way to Eureka and the coast if searching for a specific food source. Its ability to quickly and efficiently move through the forests and meadows makes its general living area a vast region.

Michelle McCardie

Retired School Teacher
Hoopa

Leanne McCardie Estrada

Social Worker
Hoopa

Sighting #5: July or August 1972

During my time in Hoopa I had been going from sighting to sighting, witness to witness and location to location with additional time spent in the far-off reaches of the reservation and national forest. There had been little time for follow-up work with governmental agencies or making contacts with key figures in the community.

It was an afternoon in early December 2006 when I found that fish weren't in the river, restaurants were closed and it was time to make some new contacts. I knew there were several people in the Hoopa Tribal Wildlife Management Group that had interesting stories about Bigfoot and they had not contacted me. I heard these stories second and third hand and wasn't able to confirm anything. I walked into the wildlife office for the tribe and I was immediately greeted with warm hellos, and several members of the group gathered around for a discussion. This was just the opposite feeling I had when I met the Tribal Forestry Team directly across the hall. This group asked several great questions about Bigfoot and some even invited me to come into their offices and look at pictures that their game cameras had taken in the forests. This was all leading to some great stories.

I made several contacts in the Wildlife Group and agreed that we would work together on a few projects and theories, and we set a date to meet in the future. As I was walking to my vehicle, one of the biology

Michelle McCardie standing in her front room. The area over her left shoulder is where she saw Bigfoot.

technicians for the tribe followed me to my car and stated that there was a woman who lives on the reservation that I should meet. He stated that she was his grandmother and she had a few fascinating stories about Bigfoot. He told me he didn't know if she would talk, but that if I would wait a few minutes he'd give her a call. I was "all ears." After a wait of several minutes and meeting a few more of the Wildlife Management Team, my original contact came back and stated that his grandmother wanted to meet me.

I followed my new friend to the far southern end of the reservation, where we took Schumacher Road to the west and then followed it further south as we paralleled the main highway. We took a small dirt road a few hundred yards and ended up at an old house surrounded by 5–10 acres of open land. As I was driving up to the house, I immediately noticed many large berry bushes that were surrounding it adjacent to the highway, and the beautiful green meadow that was nearby. The house sat probably 200 yards from the base of the mountains that surround Hoopa on the west. These mountains go directly to the coast, a trip of close to 30 miles.

I was escorted into the residence and was initially met by Lee McCardie, the 92-year-old grandfather of my escort. Lee is a retired Humboldt County Marshall and is also the former constable for the region. He is a frail man, hard of hearing, but very alert and full of interesting stories about the region. His grandson told Lee that I was there to interview his grandmother. Lee walked to the backroom and brought Michelle to the front room for me to meet.

Michelle McCardie is a 74-year-old former Hoopa first and third grade school teacher. She was also the librarian for the school for many years. Michelle is of Irish, German and Swedish heritage. She spent much of her youth living in Long Island, New York. In 1949 she attended the prestigious Simmons College in Boston and got her bachelor's degree in home economics. Michelle met her first husband when she was in high school. She explained that while she was attending Simmons, her husband was going to Harvard University. After graduating from Harvard he went onto the University of Buffalo Medical School where he interned and became a doctor. She explained that he wanted to practice medicine and legally avoid the draft. Michelle said that he opted for public service and joined an Indian hospital program, and this took them to Hoopa where he worked in the tribal hospital.

Michelle stated that in the mid 1960s she divorced her husband and met Lee. They have lived on the same property and in the same house for much of the last forty years. Michelle said that she and Lee had two children and they all lived in the same house we were presently in, on Schumacher Road. Back in the 1970s there weren't as many residences, traffic or people and Michelle said it was a much quieter existence. I should state here that Michelle is a very smart, alert and sensible person. She is one of the people I have met in the area with the most formal education coupled with a high degree of common sense. With her east coast education and upbringing Michelle almost seems out of place in the middle of Hoopa.

Michelle explained that her Bigfoot experience started in either July or August of 1972. She told me that she had a good friend who lived in an area above their house on Cherry Flats. We'll call her friend "Mary." Mary had a few clotheslines strung in her backyard that were made of heavy rope and tied securely to the side of her house on a large post. Over the course of several days something was coming into the yard and purposely breaking the lines. It was an unusual act because the ropes were very strong and this wasn't a neighborhood where kids would pull pranks. Each time the rope broke Mary would replace it. Mary was also raising several puppies in a small box that was sitting next to their house. Mary had told Michelle that the puppies were still a little young to wander the yard on their own, so she kept them confined until they had grown. One morning Mary woke up and went outside to check on the puppies and found that the box and the puppies had disappeared. Nobody had any idea who would've taken the cute puppies. It's important to remember that Mary's residence was up the hill from Michelle and was even more isolated and away from public traffic. The idea that someone would accidentally stumble into Mary's yard was absurd.

During the period that the puppies went missing, Michelle's son, Jamie, had the job of ensuring that all of their pigs in the backyard were well cared for, had food, water and that they didn't disappear from their pen. Jamie was approximately 12 years old at the time and lived in a basement room to the far rear and side of the residence. Jamie's bedroom window looked out at ground level directly into the pigs' pen. One night at approximately midnight (during that 10-day period the puppies went missing) Jamie heard a large commotion in the pigpen. The noise and disturbance was so loud that it caused him to wake up, get out of bed and make his way to his bedroom window. He pushed back his drape and was looking eye to eye with an adolescent Bigfoot. He was so scared that he screamed and ran into Michelle and Lee's bedroom. Michelle stated that Jamie explained what he saw and said that he wouldn't leave their room. Jamie spent the remainder of the night sleeping with Michelle and Lee. Michelle stated that Lee was such a sound sleeper that he never woke up that night, but was completely briefed about the encounter the next morning by Jamie and Michelle.

Early the next morning Jamie told his dad about the frightening incident. After the story, they all went into the side yard and looked at the area in and around the pigpen. Michelle stated that they saw large footprints that were similar to a human footprint but slightly larger with bigger toes. At the urging of a friend who had an interest in Bigfoot, they decided to cast the prints. They state that during the last 40 years they have misplaced the prints and have no idea where the cast might be.

During the same week as Jamie's sighting from the basement, Michelle recounted that she had a front-yard sleepover for her and Mary's kids. She explained that the kids had put tents on the grass in front of their house. It was a warm summer night and there was never a threat of any animals in the area. In the middle of the night, all of the kids woke up and heard something walking around the perimeter of their tent. The kids were too afraid to look and see what it was, but they were sure that it wasn't a bear, mountain lion, etc. The next morning the kids explained to Michelle that they were bothered during the night by something, but they were not sure what it was. Michelle said that as she walked the area looking for tracks, there was one specific area, maybe three feet by three feet, near the planter box that had a horribly bad odor. Michelle stated that it smelled much worse than any skunk she had ever encountered. The odor was confined to one small area close to the tents and it wasn't there prior to the sleepover. Michelle stated that the odor stayed in that area for several days after the sleepover incident.

A few days after the sleepover, Mary and her kids woke one morning and went into their backyard to check on several other full-grown dogs they owned. Mary told Michelle that she walked into the backyard and noticed that the box with the puppies was back at its location next to her house. Mary told her that she went over and checked on the box and found all of the puppies inside, dead. There were no obvious signs of death and it appeared that the puppies either starved or died from lack of water. They were all in the same box; same newspaper coverings and nothing specific seemed to have harmed them. It was a very puzzling incident at the time.

A few days after the puppies were returned, Michelle and Mary's kids were playing in Michelle's front yard. Michelle explained that it was in the early afternoon, a typical summer play day for the kids. Michelle was in the house when she heard the kids yelling for her. She stated that she ran out to her front porch and saw the kids pointing to the front area of the yard adjacent to the roadway. Next to their cement house she saw an adolescent-size (5 feet to 5 feet, 2 inches) Bigfoot standing next to their berry bush calmly picking and eating the berries with its fingers. Michelle stated that the Bigfoot had hair/fur covering its entire body except for its face. The facial tone was close to that of a medium complexion Hispanic person. The arms were longer than a human's arms and its face was more

Looking from the street towards Michelle's residence. The cement shed is on the left. At the time of the sighting the area was filled with berry bushes.

human than animal. She stated that the face did not have a snout like a bear, but was formed more like a human's. Michelle explained that when the creature turned to look at her and the kids, its entire upper body turned and looked. It did not appear to have a neck. Michelle could not determine a sex for the creature. The creature pulled the berries off the bush like a small child, gently but firmly. Michelle said that after the creature saw her, it calmly left the area, not running, but walking slowly. Its arms were not straight when it walked but they bent slightly at the elbow like a human's. Michelle added that the head of the creature seemed to be round at the top and not pointed.

Michelle said that she watched the creature in her front yard slowly walk out of sight and up towards the hills behind her house and up into the area of Mary's residence.

She said that this was the lone time that she actually saw a Bigfoot.

Since 1972 Michelle stated that she has heard five Bigfoot screams immediately in the area of her residence. She explained that these screams are quite different than those of a mountain lion, peacock or a young child. These screams are excruciatingly loud and come from the area on the other side of the Trinity River.

The area that Michelle indicates that she has heard the screams is very close in proximity to the screams that Hank Masten and Corky Van Pelt heard when they were on the Trinity River camping. I reiterate that these three people do not know each other and have never been told of each other's encounters.

During this point in the interview I told Michelle that it appears that she remembers this incident like it happened yesterday. She stated that it was something that she will never forget, and it was a significant event in her life. I told Michelle that I know people will read this story and they will say that she saw a bear, not Bigfoot. Michelle retorts that she has seen hundreds of bears in and around her yard. As recently as the previous week they had a bear get into their garage that they had to scare away. Michelle was absolutely adamant. "The creature I saw in my front yard eating berries in 1972 was no bear. It didn't have a snout like a bear. It picked berries off the vine with its hands and fingers like a human. It didn't walk like a bear and it was physically quite different from a bear. It was much closer to human than any bear. It looked like a cross between an ape and human. I saw a young Bigfoot."

Michelle stated that she has seen Bigfoot just one time, but others in the area had seen it after her initial encounter. She explained that Shoemaker Road was a very desolate place when Bigfoot made its visitation. Today the area surrounding Shoemaker Road and the road itself is lined with residences and a new home development. Michelle said those weeks after her Bigfoot visit she ran into people who would walk Shoemaker Road in the early morning hours and sometimes see Bigfoot in the forests and road area.

The early morning encounters of Bigfoot along the Shoemaker Road area match the time and location of the sighting by Ray's Supermarket manager (Mularkey) on his way into work.

During the entire interview with Michelle, her grandson was walking in and out of the room listening to parts of the story and helping Lee with chores. As I was leaving the residence, he approached and thanked me for spending the time dragging out an extraordinary story. He told me that throughout his entire life he had heard bits and pieces of different aspects of the encounter, but had never heard everything and had never heard how it all intertwined. He did tell me that before Michelle's son Jamie died, he had several conversations with him about Bigfoot. He distinctly remembers Jamie stating that he had seen Bigfoot looking in his bedroom window the night the pigs were making noise and that Jamie was positive it was a Bigfoot.

I was able to finally contact Leanne (Michelle's daughter and also one of the kids in the yard who witnessed Bigfoot) by phone. She is a social worker for the Hoopa tribe and lives in the community. She had spent a number of years in Los Angeles County as a social worker and is a very intelligent and aware person. She did confirm what her mom stated about the incident in their front yard. Exact details of the incident were a little vague for Leanne, but she was quite helpful in confirming the incident and confirming the Bigfoot sighting. Leanne patiently reviewed her mother's statement regarding the incident in the yard and then later signed an affidavit.

Michelle signed an affidavit on her Bigfoot sighting.

Location of Sighting

Shoemaker Road is in a prime area for making a Bigfoot sighting. Hank Masten, Corky Van Pelt and Michael Mularkey (Ray's Store manager) have all had either Bigfoot incidents or Bigfoot sightings in that immediate area; many of the sightings center on Camel Creek and the adjacent area. Michelle has told us that sightings occurred after her incidents ended. The sightings were by people walking, jogging and hiking in the early morning hours, and that matches with the Michael Mularkey sighting. This location is at the far southern end of the reservation and is also an area that has seen a significant increase of population over the years. It does appear that as the population has increased, the sightings have decreased. The only sighting or incident in recent times was Michael Mularkey's.

Michelle McCardie's Bigfoot sketches.

Forensic Sketch

Michelle is someone who has stuck very tightly to her story. She is somewhat unusual to the sighting list because she is not from the area, is not a tribal member, and is a professional with formal education from the east coast. She is a lady with an enormous amount of credibility.

The week that Harvey and Gina Pratt came to Hoopa I had known that Michelle was just being released from the hospital, and I knew that she wasn't feeling well. I made several trips to her house to see her, and each time she was asleep or resting. I scheduled several tentative dates and times for her to meet Harvey, and each time she had to cancel because of her health. Each time, though, she always said to keep calling and hopefully she would feel well enough soon to assist with the drawing. As we neared the end of the week it became obvious that Michelle would not be well enough to travel anywhere. I called her on Harvey's last day in town and asked if I could bring him to her residence and he would do the sketch at her bedside if necessary. She agreed that it would be okay.

I escorted Gina and Harvey to Michelle's residence on Shoemaker Road. We stopped in the front yard and I showed the Pratts the various locations where the Bigfoot was seen and the general layout of the area. As we walked up and into the residence, Michelle was getting situated in a large chair in the front room. She was in obvious discomfort, but was still insistent on doing the sketch. After all of the introductions were completed, Harvey got right to work.

I left Harvey and Gina at the residence with Michelle because the area was very crowded. I did not want to distract Michelle from concentrating on the drawing, and I also didn't want to make anyone feel cramped. I had spent a significant amount of time with Michelle and her grandson on previous occasions and felt very comfortable leaving Harvey and Gina at her residence.

After completing the sketch, Harvey gave me a call. He stated that Michelle was in obvious discomfort, but provided a valuable Bigfoot tool. She took her time in explaining what she saw. Harvey said that she stated that she felt that the creature was a younger Bigfoot and maybe a female based on the "sweet" face that Michelle had described. Harvey also told me that it was obvious from the description that the Bigfoot was not fully grown, but yet felt comfortable enough to visit the residence on several occasions. Harvey sated that he felt the drawing was a very accurate rendition of what Michelle had witnessed based on her response to his drawing.

Harvey informed me that he had completed two drawings of the creature. He said that Michelle had a good look at the creature's face and body structure and he felt it was complete enough to draw just a facial rendering and then follow that with a full body drawing.

Jackie Martins
(with Julie McCovey and Georgia Campbell)
Language Teacher
Hoopa

Sightings #6: Summer 1975
#1: 1950s

During one of my many trips to Hoopa I was contacted by the local elementary school and asked if I could make a presentation to the sixth grade class. One of the teachers at the school who had a lifelong interest in Bigfoot and had spent a lot of time exploring, searching and looking into the phenomena had requested the lecture. I went to the class and made a two-hour presentation on Bigfoot in their area, the science behind what has been found and what mainstream science would need to validate the existence of the creature. The class was full of great questions and a lot of "what ifs." Near the conclusion of the presentation, the teacher told me that I might want to make contact with another teacher at the school who had a personal sighting of Bigfoot. He said that she didn't talk about the encounter with outsiders, but he would try to arrange the introduction. He promised to call me at my hotel room that night.

At approximately 9:00 p.m. I received a call from the sixth grade teacher. He told me that I made an excellent presentation that he felt was filled with credible information that was fascinating to his class. He asked if I would take him for an excursion into his reservation so he could experience first hand exactly how I searched the area. I said it would be my pleasure to take him with me. He then went on to tell me that he had spoken to fellow teacher Jackie Martins, had explained who I was and what I was doing on the reservation. He said that she would be happy to talk with me. He stated that she was a Hoopa language teacher at his school and was also a past councilperson of the Hoopa tribe. He felt that she was one of the most credible people I would ever meet.

The next day I went to Hoopa Elementary School at approximately 2:00 p.m. and walked into Jackie Martins's classroom. There were a few kids still in the room working on the chalkboard, but Jackie told me to come in and sit down. I introduced myself and explained that I would

Jackie Martins behind her desk at Hoopa Elementary School.

understand if she was reluctant to talk about the sighting in the presence of students. She told me that we could talk, and that her students were all familiar with Bigfoot.

I explained to Jackie that a fellow teacher had told me that she had a personal sighting of Bigfoot and that I was interested in her story. She hesitated momentarily and then said that it was something that she generally does not discuss. She told me that she had been elected to the Hoopa Tribal Council in 2004 after being appointed in 2002. During that time she had been quite reluctant to talk openly about Bigfoot, but heard more and more stories about the creature during her years in the public sector. She assured me that she was 100 percent positive of what she had seen and would talk to me about her encounter. According to her, there are no other possibilities about what her sighting could possibly be; she knows she was a witness to Bigfoot. Jackie said that she is a Hoopa language expert and finds it fascinating that there is a specific name for the Bigfoot creature in her language that dates back over 200 years. Back then her people were too busy trying to survive and were not prone to making up words and stories about a non-existent creature. She said that elders have kept stories about Bigfoot and its culture as a traditional part of Hoopa life.

As a Hoopa language teacher at the elementary school, she hoped that the kids would keep the traditions and language alive in future generations. She stated that she loved her job and appreciated the positive energy that the kids brought to her class and life.

Jackie said that her Bigfoot encounter occurred many years ago in the hills above Hoopa. She said that in 1975 she was 19 years old and living in Crescent City with her good friend, Julie McCovey. Jackie said that Julie was 17 years old and they both enjoyed going to dances sponsored by the tribe. It was either July or August when they both heard that there was a dance in Hoopa, and they decided to make the drive. She was driving her old red Chevrolet El Camino and Julie was the front seat passenger. They contemplated taking the paved road around Eureka and through Willow Creek and up the Lord Ellis Grade, but both agreed it would take too long. They decided to take the dirt road straight from Orick over Bald Hills. Both knew that this road wasn't heavily traveled and if they encountered a problem there wouldn't be police, tows or emergency services to assist them. But both girls grew up in a rural area and were fine with the decision, and saving more than an hour of travel time by going over Bald Hills was the added bonus.

Jackie said that when they left their house it was still sunny. It was almost 9:30 p.m. when they crested Bald Hills summit and were making their way to Martins Ferry on the Klamath River (this is located just west of Weitchpec). They were traveling downhill on a series of switchbacks on the dirt road and were approximately 10 minutes from the Klamath. About dusk they were making a turn in the road and Jackie had her high beams on. As she was clearing the turn, they both saw a huge creature come off the embankment on the right side of the road. It was walking upright on two feet, had hair over its entire body and had the classic Bigfoot appearance now familiar in videos. They were approximately 200 feet from the creature when the beams of their headlights hit it. They saw the creature turn its entire upper body (not just its head) to look at their vehicle. As the creature looked at them, its pace accelerated towards the other side of the roadway. She described the creature as walking with its knees slightly bent and always walking on two feet, very similar to a human. Jackie said that the creature's hands hung down near its knees as it walked. Her best guess is that the creature was near seven feet tall and wasn't skinny or fat, just well proportioned with broad shoulders.

Jackie's recollection was that the creature had heavily textured hair or fur, as she could see it moving in her headlights. She did not see the creature's face clearly enough to identify it.

Jackie told me that she had seen more than 20 bears in her lifetime and had seen several stand on their hind legs. She was sure when she looked at this creature that she was looking at Bigfoot and not a bear — 100 percent positive. She said that the creature didn't move like a bear, it moved much like a human. It looked like it was meant to be walking on its two legs.

Ms. Martins went on to say that as she and Julie were looking at the creature, Julie was screaming with fright, begging her to back the car up and get out of the area, but she didn't. Jackie drove forward as the creature went down the embankment. Both girls were very concerned that the creature would catch them at the next bend in the road, but it didn't. Jackie said that they never saw the creature again. She estimated that the total time she was able to observe the creature was 5–7 seconds and it appeared to her that the creature was afraid of them. She considered the sighting a blessing, while Julie felt as though they just had seen the Indian Devil. Jackie said that she is a Hoopa tribal member and Julie is a Yurok tribal member and that at some level that may account for the interpretations of the sighting.

The girls continued to drive into Hoopa without any other incidents. When they learned that there actually wasn't a dance that night in Hoopa they drove to Jackie's aunt, Cleo McCardie's house. The girls told Cleo about their sighting and it was Jackie's feeling that Cleo believed her. They spent the night at Cleo's house and awoke the next morning to have another aunt, Georgia Campbell, visit. Georgia was 50 at the time and Cleo was slightly older.

Jackie said that Georgia took her off to another bedroom and told her that Cleo had told her of the Bigfoot sighting the night before. Georgia revealed that she didn't tell people, but because Jackie was a witness of the creature, she would tell her. Georgia said that she had seen two Bigfoot in her lifetime. When she was living at Bull Ranch (in the eastern hills between Hoopa and Weitchpec) she was working in her kitchen late one afternoon. She was looking out her front window when she saw what she believed to be a mother Bigfoot and an adolescent Bigfoot crossing the field in front of her house. Georgia told Jackie that she never told anyone this story because she didn't feel that people would believe her. She said that since Jackie had seen the creature and knew that it was real, perhaps she would believe her story. Jackie said that she totally believed Georgia's story. Jackie told me that both Cleo and Georgia are now deceased.

I contacted Julie by phone at her residence in Eureka. We spent considerable time talking as she recounted the incident with Jackie. Julie confirmed everything that Jackie stated and promised to sign an affidavit and swear to the events as they happened. Unfortunately, I could never get together with Julie to get her to sign the affidavit. However, Jackie did sign an affidavit.

Location of Sighting

The location in the Bald Hills of this sighting is very significant. This is an area with a history of Bigfoot sightings and is a region where I have uncovered many Bigfoot related incidents. It appears to be a byway that Bigfoot uses frequently to travel between the area east of Hoopa and the coast. There have been a number of sightings and occurrences in these mountains by people who are totally unrelated and know nothing of the other sightings. Many of the sightings and incidents have occurred between tribal and non-tribal members.

An interesting aspect to many of the sightings in this area is that the coast is involved in one way or another. Either people are traveling to or from the coast, they are looking for people coming from the coast, or they speak of the coast in their sighting report. This area is very lush, with lots of water and thick vegetation. I have spent considerable time on these mountains and I have never seen a deer. This may be indicative of the thick vegetation, but I doubt it. There are many homes on the Hoopa side of the mountain, but homes are scarce as you move inland. I have seen many bears in this area and parts of the mountain are thick with berries and other edible plant life. The Trinity River is at the base of one side of the mountain and the Klamath River is on the Martin Ferry side.

I normally don't include sighting reports where an affidavit is not signed, but an exception was made in the case of Bull Ranch. Jackie's story of a relative coming forward and talking to her about her personal story parallels many stories I have heard about people being reluctant and embarrassed to talk about their sightings, which aren't in this book. Georgia Campbell's sighting on Bull Ranch fits completely into the idea of Bigfoot living in the mountains east of Hoopa and migrating west towards the coast utilizing the Bald Hills. The area of Bull Ranch is fairly isolated with very little vehicular traffic, but it is also adjacent to Hopkins Butte, supposedly an area that the Hoopa elders tell others to stay away from. There is only one small house in the area today and the people who live in it are rarely there. I know the people who live in this house and they are aware of Bigfoot, but haven't seen one in that area or had any unusual incidents. Bull Ranch is approximately halfway between Hoopa and the Bluff Creek Resort and is in the middle of the region where many elders believe that Bigfoot resides today. The entire region north of Bull Ranch is an area that Hoopa people do not travel, and believe this is a sacred area for Bigfoot. The Bull Ranch site would be on the sunny side of the mountains and Bald Hills would be the shady side. It would be a fairly short walk for Bigfoot to migrate to the area of the Klamath River, Weitchpec and the area behind Bluff Creek Resort if they were moving from Bull Ranch.

Jackie Martin's sketch of the Bigfoot she saw on Bald Hills Road.

An interesting story about the Bald Hills area involved me one warm summer afternoon. I was making my way from Hoopa to Redwood National Park when I was near the summit. I had stopped and was attempting to decipher on maps exactly where I was. As I was parked in my jeep in the middle of nowhere (with my "California Bigfoot Search" stickers on my car doors) a beautiful woman on a horse came down the middle of the road with her dogs. She rode up next to me and asked if I needed assistance. She did help me with the directions, but this led to questioning her about Bigfoot. She stated that her family owned a large ranch in the area and they had lived there for many years. She explained that they never had any direct Bigfoot sightings, but they had a very unusual event happen several months before. They had a very large garbage can that probably weighed 100 pounds at the back of their residence. She explained that they had heard a noise in the middle of the night, but never got up to check, figuring it was a bear. The next morning they went out to dump the garbage and found the garbage can was gone. Nothing was dumped; the garbage can and garbage were gone. She stated that her dad and brother searched a radius of 300 yards and could find anything. The family thought that Bigfoot might have taken the can and its contents, but they never found out the truth. They never have found the can.

I told the equestrian I appreciated the directions and the story, and preceded down to Redwood National park. I also left her my card and asked her to contact me if there were any new developments.

Forensic Sketch

Jackie was very enthusiastic in working with our artist on the sketch. From the first time we met, she offered to take a polygraph test should I, or anyone I worked with, not believe her story. I told Jackie that I felt she was a very credible person and there was absolutely no need to polygraph her.

Jackie started her meeting with Harvey by describing the creature that she observed on Bald Hills Road. She said that she felt it was seven feet tall, had hair on almost its entire body and was dark in color. She also stated that she felt its hair was 3–4 inches long. She explained to Harvey that the creature had arms longer than the human arm, and they hung down to near its knees. She stated that she felt it was a male and that it was in very good health. She couldn't offer any more of a description of the creature's face other than what Harvey drew because of the lighting conditions and the distances involved.

Tane Pai-Wik at her residence.

After Harvey had completed the drawing, Jackie was very impressed with the results. She stated that the drawing was an excellent representation of the creature that she and her friend saw that night on Bald Hills Road.

Tane Pai-Wik

Mother
Hoopa

Sighting #7: Summer 1984

If there was one person in Hoopa that I had heard many stories about and was told that I definitely needed to interview, it was Corky Van Pelt. It took many months before I found someone who told me where Corky's mom, Gwen, lived. I went to her house, explained my interest in meeting

Tane's drawing of the creature she observed.

Corky and asked if she could introduce us. She said that Corky did not have a phone, but lived up Bald Hills Road near Transmitter Hill in a very isolated area with his wife and kids. Gwen drew me a very complicated map and said to ask neighbors if I got lost — good idea!

My first trip to locate Corky I stopped and asked three sets of neighbors for directions. All of the people I met were very nice and helpful, and eventually I made my way to his residence. On my first trip I met Tane, Corky's wife.

Tane was very polite and invited me into their home. She was babysitting her three kids and it was fairly early on a weekday morning. Tane stated that Corky had gone into town to see a doctor about his bad back. I explained to Tane that I was there to talk to Corky about his Bigfoot encounters and hear his stories. Tane now politely asked if I had the time to hear her Bigfoot stories, and I replied that I absolutely did have the time.

Tane said that she was a Yurok tribal member and grew up in the area and attended Del Norte High School. She earned her high school equivalency degree and never went to college. Tane was well spoken and had a calm, confident sense about her. She said that she met Corky at a "Brush" dance and they had been married for almost 10 years.

Tane stated that the sighting has never left her and it is something that will stay with her forever. She told me that when she was in high school she did an art project where she drew the creature she witnessed. She allowed me to photograph the drawing.

As I was talking to Tane, I was looking around her room thinking about how ironic it was that I was at the same house months earlier regarding another Bigfoot story with Warrior Sanchez. I was later told that Warrior moved out of the residence and back into the city, and Tane and Corky moved in. Tane never mentioned anything about Warrior, and I didn't want to detract from why I was there.

On or about the summer of 1984, Tane was five years old. She stated that she would spend time that summer with her aunt, Marion Frye, and her cousin. The family had a small cabin in Pecwan, west of Weitchpec along the Klamath River. Tane said that the cabin was a wreck and could almost be considered a shack. The cabin was sitting far off the main road and had very little traffic or people frequenting the area. Tane said that she was so young that at night she would sleep with her aunt in a bed that was pushed up against an open window. It was still warm enough at night that there was no glass, only a black garbage bag that partially covered the window frame.

Tane said that she spent many days and nights at the house in Pecwan and never remembered anything odd happening until early one morning when she and her aunt were lying in bed. A horrendous odor filled the air and it seemed to be coming from the exterior of the cabin. At the same time she and her aunt heard footsteps outside that seemed to be coming close to the window. Tane said that it was at this point that her aunt forcefully told her to hide her head under the blankets. Tane said that she initially did what her aunt told her to do, but, as any kid might do, she had to look. She looked out the window and saw a huge creature's head enter the window. The creature had dark skin, hair on its face, a huge flat nose, and it appeared to be half ape and half man. The creature had a very high forehead, round eyes and had leaves and dirt on its matted hair on its face and head, as if it had just woken up from lying on the ground. Tane said that the creature kept looking in the window at them until her aunt yelled something at it in Yurok. The creature made a grunting type sound and walked away from the window.

Harvey Pratt's sketch of the Tane sighting.

Tane said it was years later that she saw a television special on Bigfoot and it reminded her of what she saw in the bedroom that morning. She said that the image of the creature is frozen in her memory and she will never forget. She is positive that she and her aunt saw Bigfoot that morning.

Tane said that she has been back to the shack and looked at the ground outside the window where she was sleeping. She stated that from the ground to the bottom of the window is approximately six and a half feet and the creature's face was centered in the window, thus meaning that the creature had to be over seven feet tall. She said that she never has seen Bigfoot other than the one time with her aunt. She also said that she has only smelled the same putrid odor two times in her life. The first time was in the bedroom with her aunt, and the second time was outside of the residence where she now lives. Tane told me that the odor is nothing like a smelly bear or skunk; it is different and much, much worse.

Tane said that the moment in time when Bigfoot had its head in the window at the cabin is something she can't and won't forget. She takes it as a blessing that she saw the creature and that it obviously meant them no harm. Tane invited me back the following day to meet with Corky. Tane signed an affidavit.

Forensic Sketch

In March 2007 Tane agreed to meet with our sketch artist and describe her sighting. In the pre-interview with the artist, Tane described the creature as an older male with thinning hair on its head and a very large forehead. She stated that the teeth were old, green and bordered on rotten; she did not see any canines. She remembers seeing the veins in the creature's eyes and described the color of the eyes as brown or very dark green and appeared bulgy. She stated that the creature definitely wasn't human and she has always felt that it may have stayed in the cabin in the past and was upset that they were now in it. Tane explained that the cabin was very old, not really secure and anyone could have been in it. She told the artist that the creature had some type of vegetation in its hair and it appeared fairly unkempt and dirty.

Harvey Pratt took approximately 90 minutes to complete his drawing. At the conclusion, Tane felt the drawing was a very good representation of what she saw. If you compare what Harvey drew and what Tane drew, there are similarities. Harvey, Gina and I agreed that this sketch showed some facial human similarities, but Tane was positive that it wasn't completely human.

Location of Sighting

Pecwan is located at the end of a dead-end stretch of highway that begins in Weitchpec. This is a very desolate area of the reservation, immediately adjacent to the Klamath River in an area known for years as Bigfoot country. There is no way to get to the coast by road in this region other than backtracking through Weitchpec and Hoopa, or traveling over Bald Hills. This is also the region that several Humboldt County deputies told me was lawless and dangerous, and not to enter it unarmed. The area has been known for years as an area for illicit drug labs and growing marijuana. I've never gone into the area with a tribe member, and would recommend the same for anyone having an interest. Stay on the Bald Hills side of the Klamath for conducting research in this area.

Jan Wyatt and Alice Barker

Retired Fire Lookout/U.S.F.S.
Gasquet

Sighting #8: 1984/85

I've always felt that receiving sightings reports and meeting certain people is dcstiny. Thc day that I mct Jan Wyatt was onc of thc days that portended destiny.

In April 2007 I had my kids during spring break. We had talked about several possibilities, but both kids said they wanted to understand how I met Bigfoot witnesses, how I tracked Bigfoot and the places it lived, and how my days in the field were spent. I worked for one week solidly developing an itinerary that would keep the kids and me busy. I had always wanted to spend time in the far northwest corner of the state along the coast. This was an area where I spent considerable time as a young adult fishing and enjoying the outdoors. We decided to leave on a Monday morning and spend the week on the Redwood Coast.

During one of our days traveling and roaming the area along Highway 1, I decided that we'd drive up Highway 199 and spend time meeting locals in the Hiouchi and Gasquet areas. I had a few anonymous sighting reports of Bigfoot in those areas and wanted to see what the true level of activity was in that region.

One of my standard practices when I enter any new community is to contact the local postmaster, especially in small communities. I met the woman behind the postal counter in Gasquet, explained my background and asked if there were any locals who had had Bigfoot encounters. She immediately stated that Jan Wyatt was a local with Bigfoot encounters and encouraged me to talk with the grocer who could supply me with her address. I walked down the street and spoke with the clerk in the market and he gladly gave me directions to Jan's residence.

My son was sleeping in the car when I arrived at Jan's so my daughter joined me in my interview. Jan was a vibrant 74-year-old widow with five grown children. She was an extremely energetic woman who would keep any 20 year old on their toes. She was very happy to meet with me and invited us into her home. Jan introduced us to her dog, Bruce, who followed us around the yard and eventually into the living room.

Jan told me that she spent a lot of time in the woods. She had worked for the United States Forest Service as a fire lookout for many years, and then did work for them as a fire engineer and other road-related work. Born in San Diego, she spent a lot of her life in the Orleans and northwestern coastal areas of California.

My ears started to perk up when she said that she had spent time in Orleans. I asked her how she ended up in Orleans. She told me that the forest service office in Smith River had sent her there to work on several occasions and she enjoyed the location. She had a sister-in-law who had been traveling in that general area and had a Bigfoot encounter, and Jan would describe it if I were interested. I told her that I was very interested.

Jan said that her sister in law, Alice Barker, was a homemaker who lived in Crescent City. Jan stated that she and Alice were close and talked often. In 1984 or 1985 Alice was traveling from the coast to Orleans and was using the shortcut over Bald Hills Road. She was with her boyfriend and they were traveling at night. Alice told her that she had to use the restroom when they were in the middle of their trip. Jan said that Alice explained that she had her boyfriend stop the car and she exited and found a place in the forest where she could urinate. Jan said that she was only slightly into her task when Bigfoot walked directly up on her. She was so scared, she ran to the car with her pants down to her knees urinating all the way to the car. Jan said that Alice described Bigfoot as a huge, two-legged creature with hair over its entire body and standing well over seven feet tall. Jan said that it was obvious that Alice was embarrassed over what had happened and was a bit sensitive about telling Jan. Jan swears that she is 100 percent positive that Alice had a Bigfoot sighting on Bald Hills Road in 1984/1985.

Jeep on Bald Hills Road.

Alice died of a heart attack when she was in her early forties. Her boyfriend also died right before Alice passed away.

Location of Sighting

Bald Hills Road starts at Martin's Ferry at the Klamath River and inside the Hoopa reservation. The road skirts through the reservation and eventually goes into Redwood National Park and terminates at Highway 101 at the coast.

Bald Hills Road is one of the most unusual roads in California. The road is partially inside the Hoopa reservation and thus they can restrict some access to it on the reservation, but most of the road is public. The middle section of the road is on property owned by the United States Forest Service. This portion of the road skirts public and private property and is a well-maintained dirt road from the Martins Ferry Bridge until you are halfway into the Redwood National Park. The road meanders its way through the mountains and then makes a descent into the park. You travel through several huge fields with miles of great visibility. You end up in the park with thick, rich vegetation that includes 2,000-year-old redwoods and huge firs. In almost every national park in the United States you would expect a restricted gate access to enter the park with a fee, but not at Redwood National Park. There are no gates coming from the east or west.

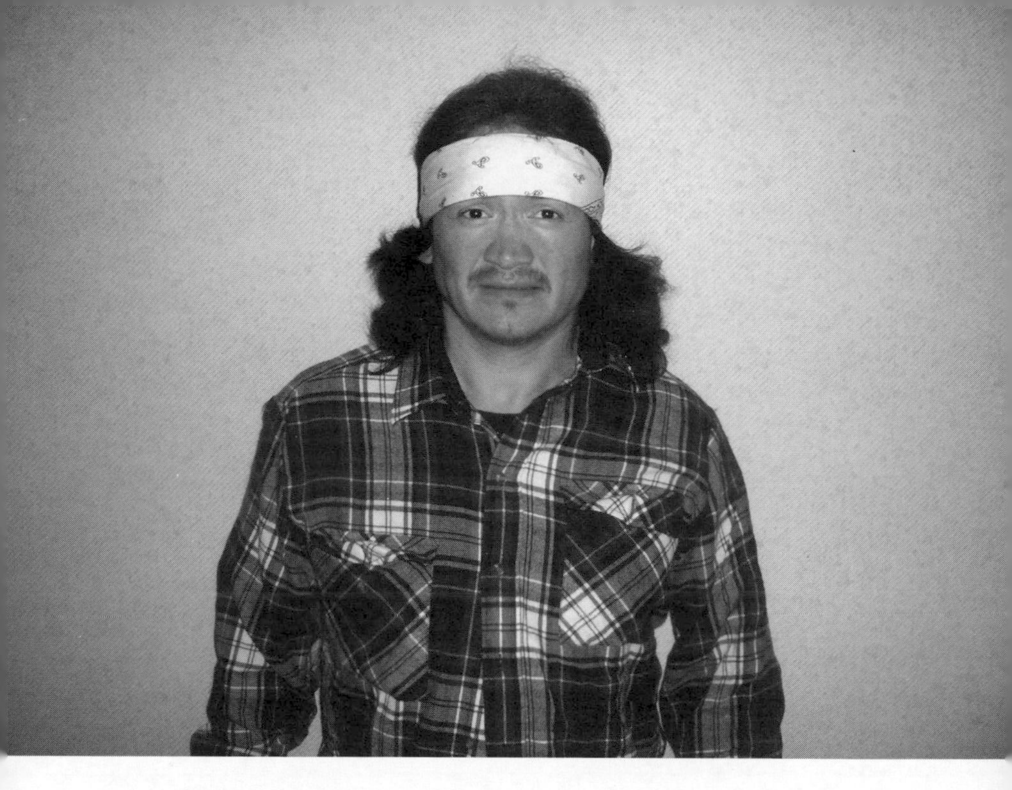

Damon Colegrove at Hoopa.

At the midpoint of the park, Bald Hills Road turns to asphalt for the remainder of its length to U.S. Highway 101. Redwood National Park and Johnson Road that forks off inside the park are two of the most active Bigfoot spots in the western United States. Once Bald Hills enters the high coastal range, sightings do not slow down. I have personally interviewed many witnesses who have seen Bigfoot on the section from the mountains down into Martins Ferry.

The sighting of Alice Barker fits well into the profile of sightings in this area. It also fits well into the Hoopa tribal area and the number of sightings in that region. Alice was not a tribal member and Jan Wyatt didn't know I had been doing extensive research into that area prior to telling me the story.

There are only two sighting reports I have included where there was no affidavit signed, this sighting and that of Georgia Campbell. Both witnesses are deceased, and both reports are secondhand accounts of the sightings. This one was too close in proximity to other documented encounters not to include it.

Damon Colegrove

Logger
Hoopa

Sighting #9: 1984

Damon is a local Hoopa resident who contacted me while I was staying at the Tswenaldin Inn. He stated that he heard that I was trustworthy and looking into Bigfoot issues and he wanted to relate a sighting that happened to him.

Damon said that he had lived in Hoopa a majority of his life and had graduated from Hoopa High in 1984. He said that he was 40 years old and was a logger. He also reported that he was a portrait artist and could draw the picture of the creature he observed.

Damon claimed that he was raised with a rifle in his hands, having been deer hunting since he was young and having spent much of his youth roaming the reservations and the surrounding hills and valleys. On almost every hunting trip he would get a deer and he felt he was a lucky hunter. He also stated that his elders had taught him to hunt quietly and efficiently, and if he utilized his skills he could get very close to the game.

In 1984 Damon was hunting an area just east of the reservation, but an area that he had known well while growing up. He had crossed the Salmon River and was hunting up Camp Three Road near Somes Bar. This is a very desolate mountainside and doesn't get a lot of hunters. It's steep and the hunting is rugged. There is a road that goes up an embankment and then levels off on the ridgelines on the eastern side of the Salmon River. He said it was very early in the morning, maybe 6:30 a.m. when he took a shot at a buck and started to track it. He was moving through chaparral-type brush when he slowly started to enter a clearing. As he reached the clearing he quickly turned a corner and he was suddenly standing under a huge oak tree. Also standing under the oak tree was a gigantic Bigfoot that was looking directly at him.

The creature was standing on two feet, had grey streaks in the hair near its beard, and had long hair, but not really thick, and the color of the hair was dark brownish-red. The facial skin under its hair was pale. It did not have a belly but did have wide shoulders. Damon estimated that the creature was eight and one half feet tall.

Elk on the roadway near Somes Bar, very close to Damon's sighting.

The creature never moved or did anything, but just looked at him. He said that creature did not have an odor that he could detect. Damon characterized the creature's face as more human than animal. He made a point of stating that when the creature looked at him, it had a look of intelligence.

Once in front of the creature, he thinks he remembers freezing solid for a few seconds. In his native tongue he told the creature not to be afraid. He then took long strides away from the creature until he got out of sight; then he ran. He told his mom about the incident when he got home, but has tried not to think about it much when he's in the woods. Damon did sign an affidavit.

It's a common belief among many of the tribes in this area that if you confront a Bigfoot that you should talk to it in your native tongue, tell it to stay calm and then slowly back your way out of the area. The Natives believe that Bigfoot understands their Hoopa language.

Location of Sighting

You must drive east on Highway 96 from Orleans and, prior to Somes Bar, take the Salmon River Road south, go one mile and take Camp Three Road east up into the hills. There is a major transmission transfer station approximately three miles up the road. The road to the transfer point is narrow, paved and in great condition, with several hairpin turns; caution should be utilized. Once into the ridge area of this road you will immediately notice how desolate it is. I was there in the middle of September on a beautiful day and never saw anyone.

The area is pristine, quiet, with lots of pine and oak, a perfect Bigfoot habitat. There are several water sources with many large trees and miles of virgin forests. This specific area hasn't had a fire in many years, which only adds to its pristine condition. There were a few small meadows that I found where you could view wildlife. It's a little strange to go from a nicely paved road to dirt and never see a vehicle. Several times I pulled to the side of the road to explore. The extreme quiet of this forest was a little strange. The truly odd encounter I had on this trip was when I was leaving. I came around a tight turn and drove up to a herd of elk standing in the middle of the road. There were probably close to 20 that I could see, and several up the hillside that were partially visible. The sighting of elk in the area confirms that there is a large viable food source that frequents the region.

After receiving Damon's report I discussed it with several of the elders in the Hoopa tribe. They told me that there are areas of the Salmon River that have been known by elders as an area that belongs to Bigfoot. There were several stories told to me by people who were told by their relatives not to travel overnight to the area east of the Salmon. Several were told not to go into the area at any time.

In analyzing the sighting and how Damon was able to walk up to the

Forensic sketch of Damon Colegrove's sighting.

Bigfoot, it fits the profile of many sightings by hunters encountering Bigfoot. The hunter shoots an animal, starts to stalk the game; at the same time Bigfoot hears the shot and starts to stalk its game. Hey, it wants an easy meal if it can get to it first. Bigfoot appears to have made a connection between the gunshot and wounded game. Damon probably hit the deer, wounded it and was close to locating it when he walked up to Bigfoot. The story is plausible because of the sequence of events and the proximity of the shooting to the sighting.

Oregon Precipitation Map.
Copyright 2007 PRISM Group, Oregon State University, www.prismclimate.org

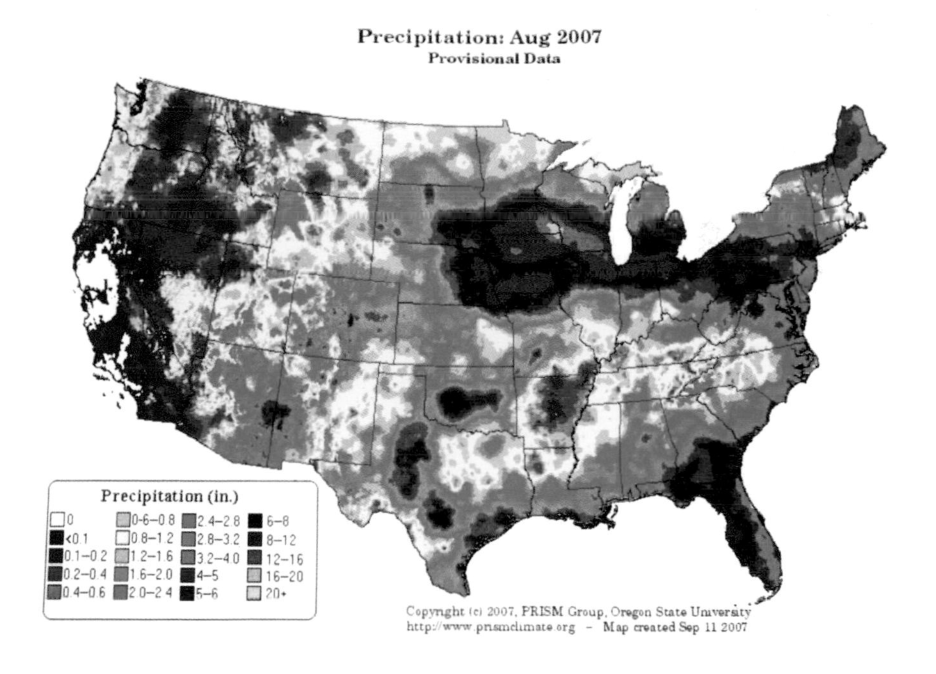

U.S. Precipitation Map.

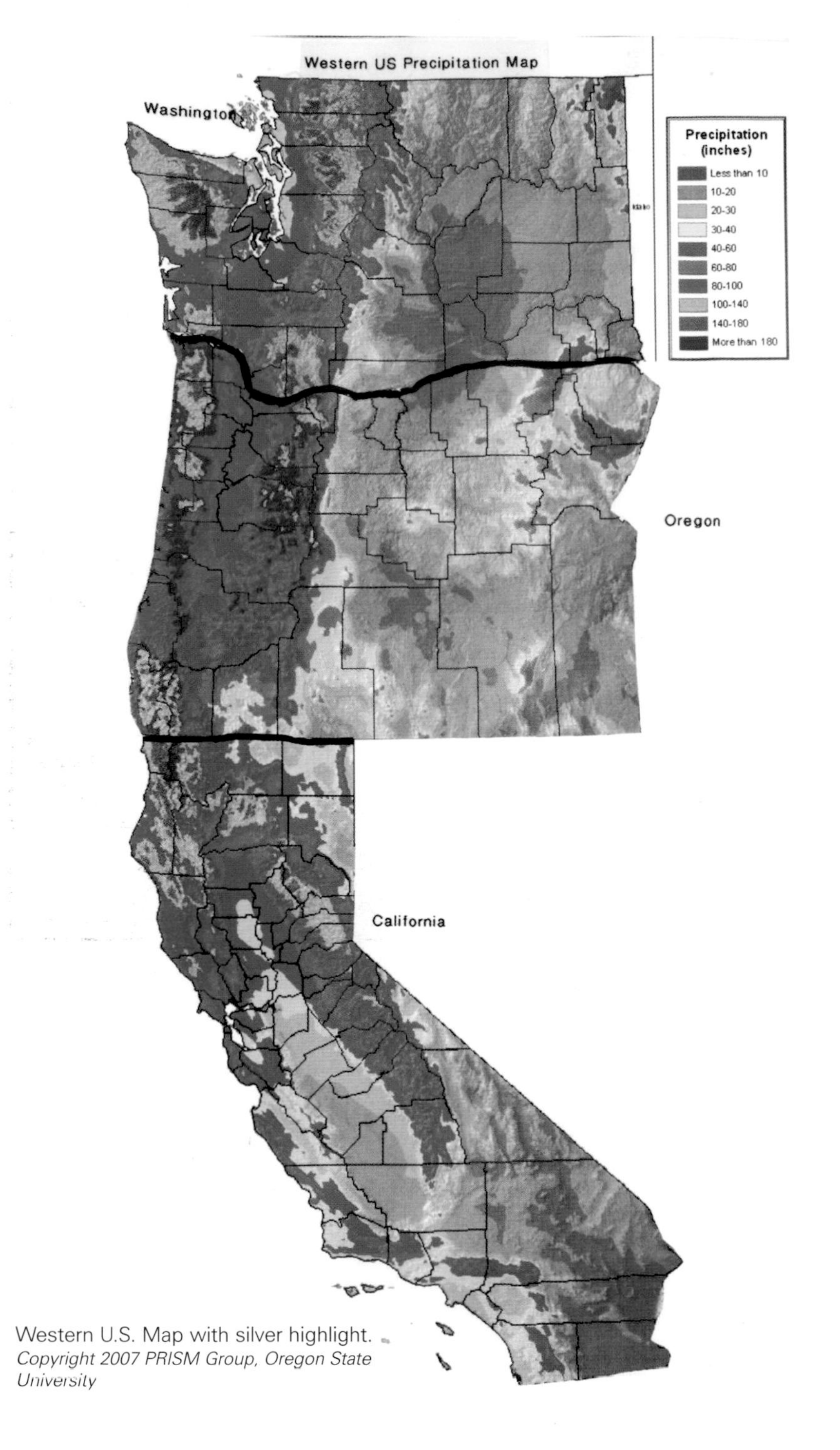

Western U.S. Map with silver highlight.
Copyright 2007 PRISM Group, Oregon State University

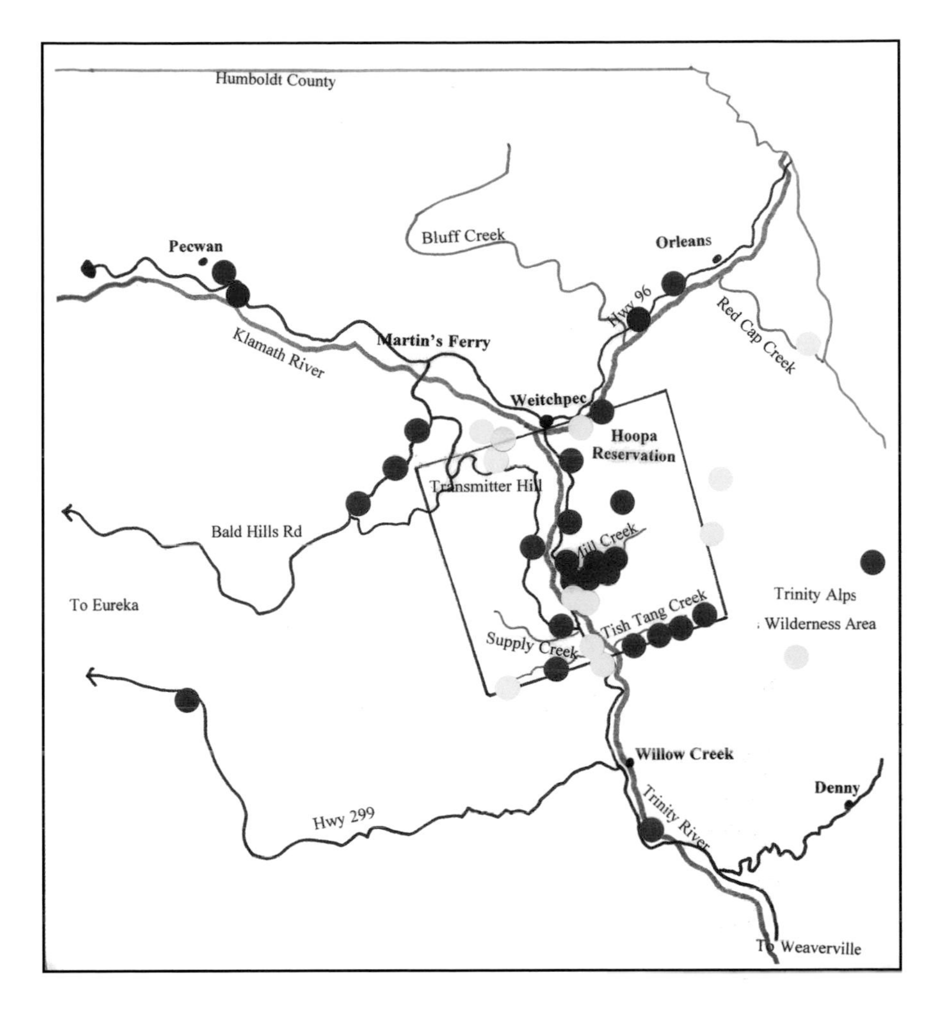

Hoopa Reservation Area Map.

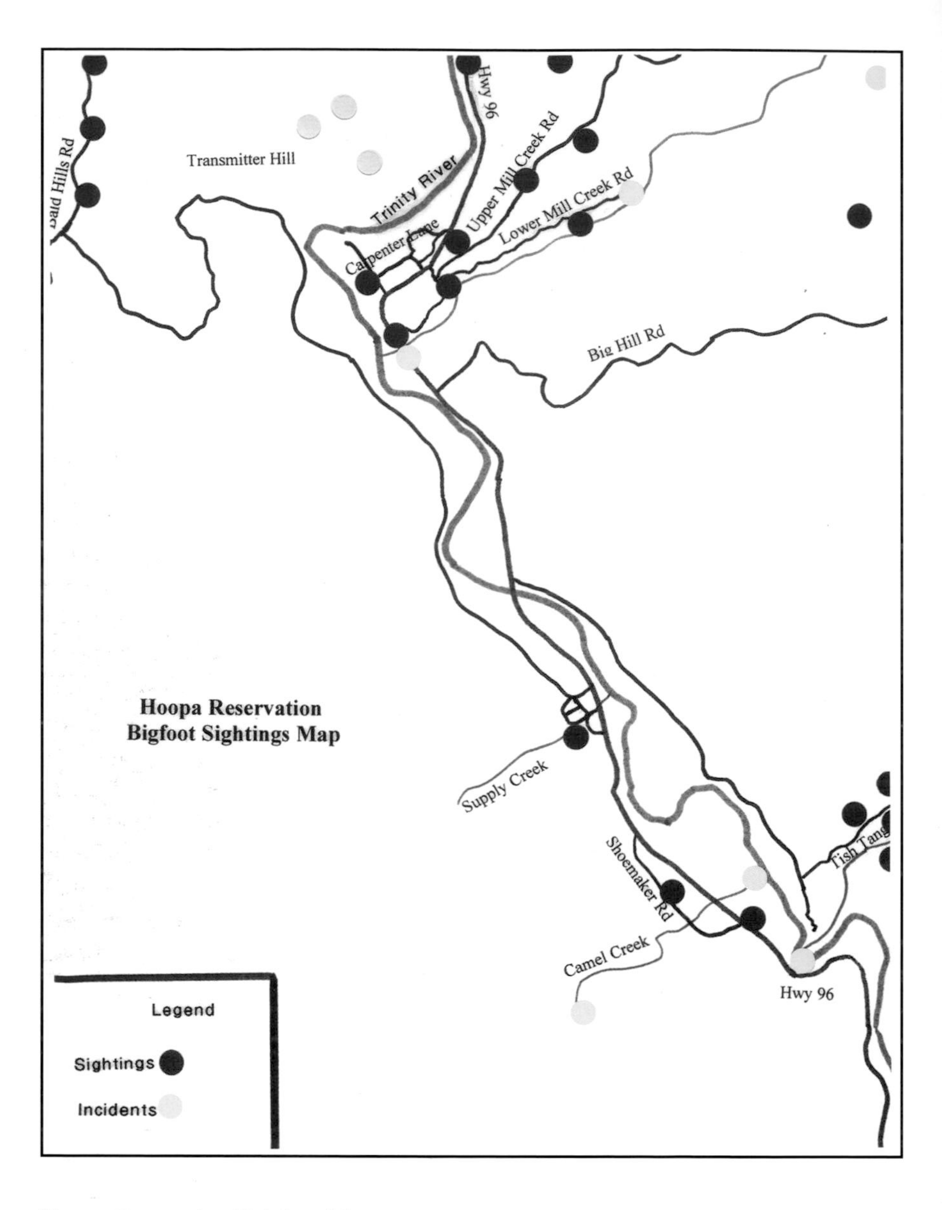

Hoopa Reservation Sightings Map.

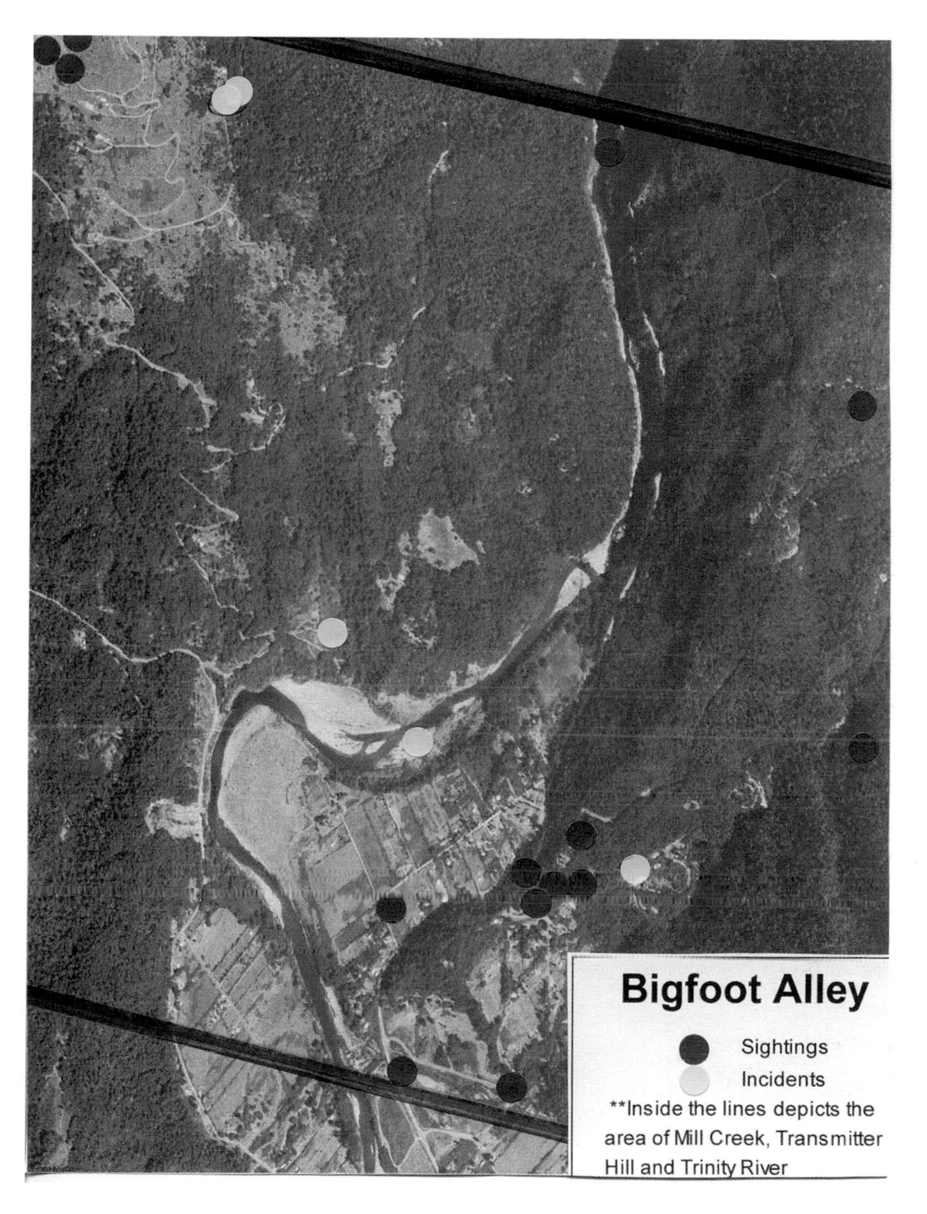

Aerial Photo of Bigfoot Alley.

Damon sketch.

Martin sketch.

Tane's drawing in high school.

Harvey's sketch for Tane.

Ed Masten sketch.

Opposite page
Mularkey's sketch.

Juliene drawing.

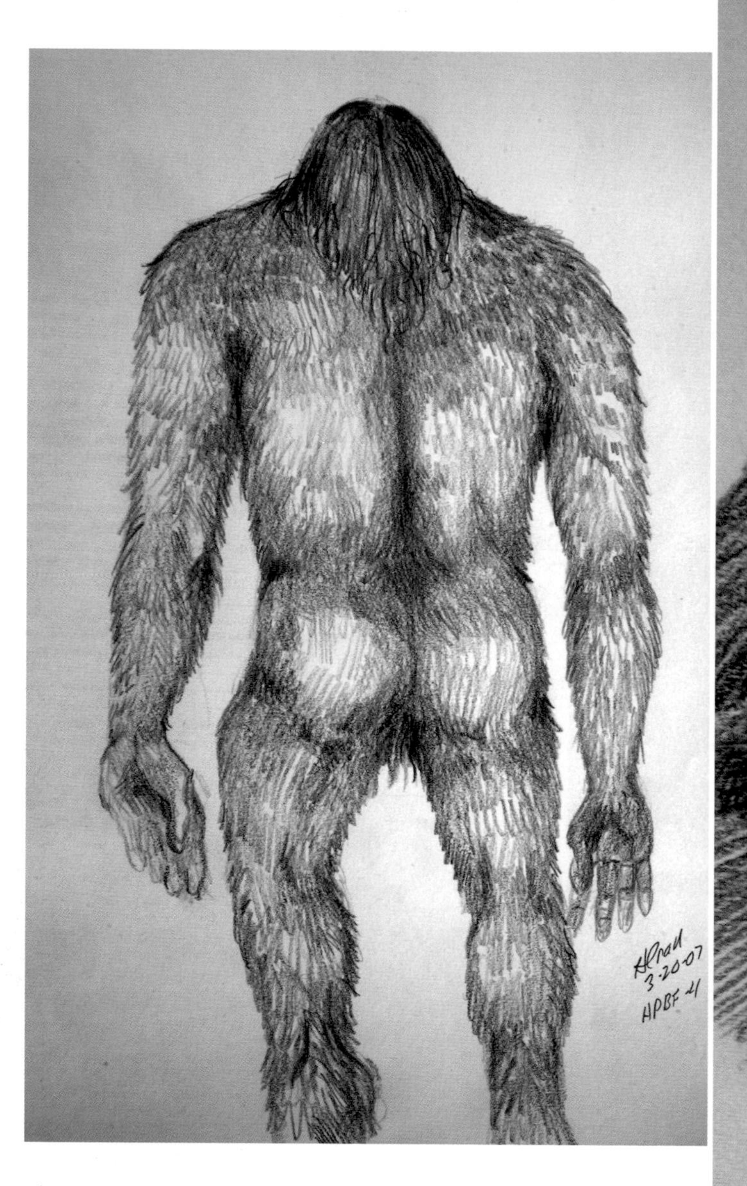

Raven back sketch and facial sketch.

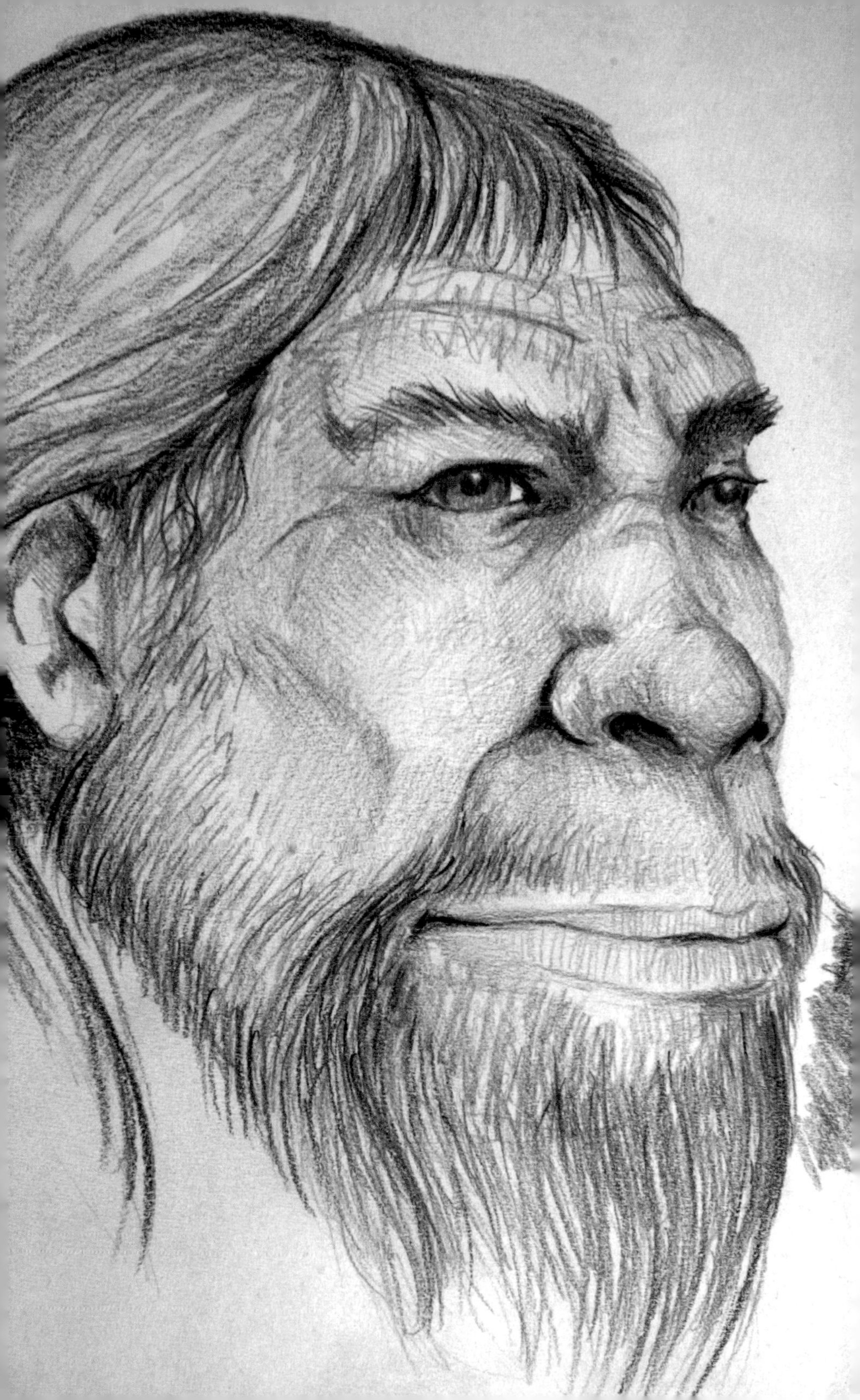

Harvey Pratt's sketch of the creature that visited Josephine.

This page and opposite page Michelle McCardie's Bigfoot sketches.

Romeo's sketch.

Above Inker's hillside sighting sketch.
Opposite page Inker's creature's arm.

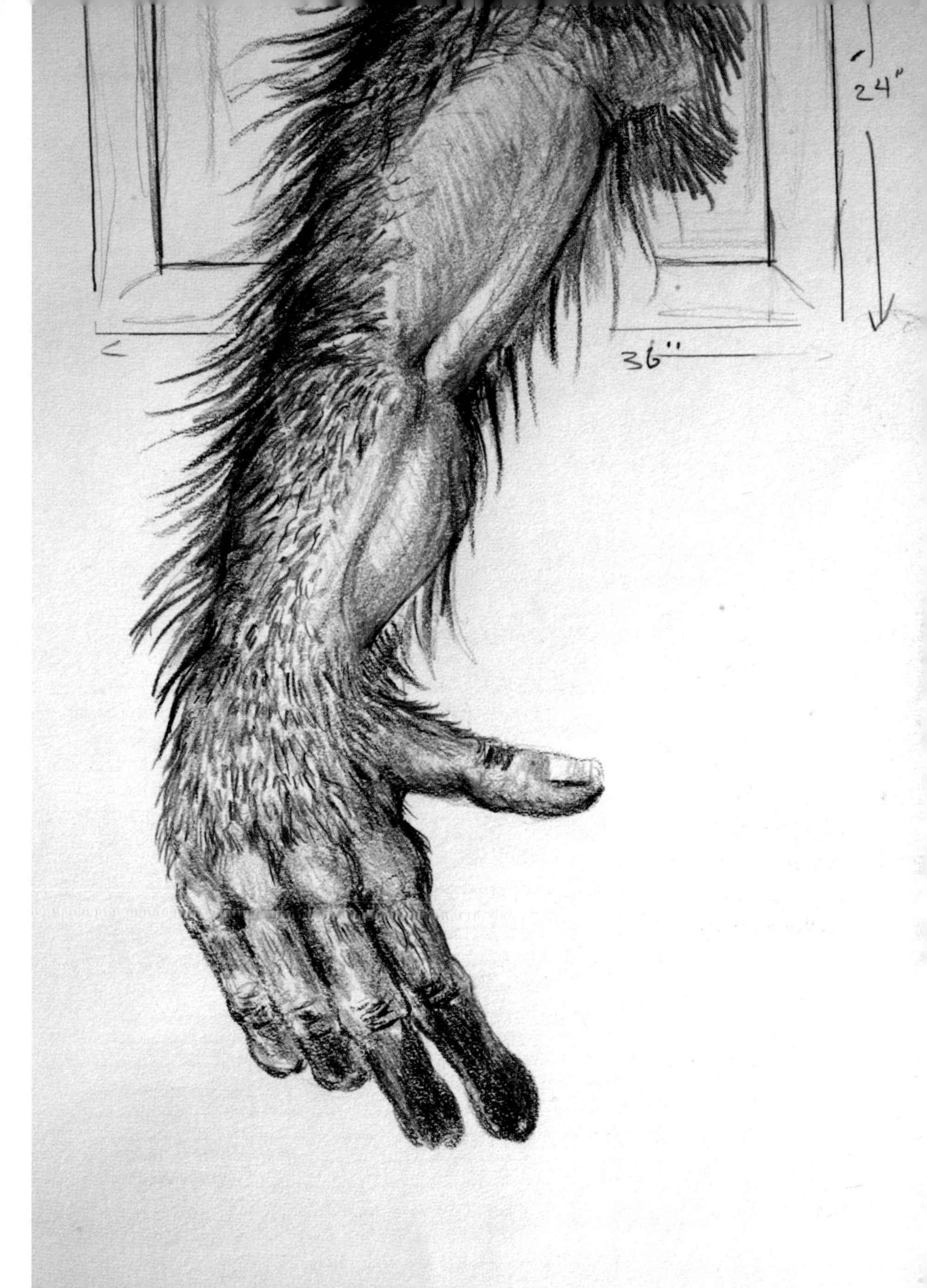

24"
36"
1964-65

Mary McClelland sketch.

Kim Peters sketch.

Damon agreed to meet with our forensic artist and describe the creature that he witnessed. He had told me early in this investigation that he was an artist and that he'd draw the creature for me, but he never did. He did however meet with Harvey and Gina Pratt and spent almost two hours describing and assisting Harvey in the sketch of the creature. Damon, like most artists, was into the detail and specifics of the sketch, which was great for the description.

Damon told Harvey that his grandfather advised him that the creatures understood Karuk. He said that he told the creature in his native language, "Don't be afraid." Damon said that the creature he confronted was absolutely not human; it was Bigfoot.

Colegrove physically described the creature as 8½–9 feet tall, 400–800 pounds with a powerful build. The creature had huge shoulders that were pointy and it had one-inch-long, gray streaks in its beard. It had a long torso and appeared in very good condition with healthy shiny hair. He stated that he didn't see any debris on the creature and that its ears stuck out somewhat, but were covered by its hair.

Damon told Harvey that as he ran away he looked back and thinks he saw the creature pointing downhill. He wasn't sure if that meant that he should continue to run away or if the creature was pointing to others this the direction that Damon ran.

Damon was very impressed with how much Harvey's sketch looked like the Bigfoot he saw. I saw Damon the following day at the supermarket and asked him what he thought of the experience. Damon stated that Harvey was one heck of an artist!

Lillian Bennett

Forks of Salmon

Sighting #10: Summer 1985

During my time in the Hoopa and Orleans region I had read and heard about a number of Bigfoot sightings in the Forks of Salmon. Approximately

Lillian Bennett with her shotgun in her residence at Forks of Salmon.

10 miles southeast of Orleans is a small paved road that follows the main fork of the Salmon River for 20 miles to the small town of Forks of Salmon. The town was established over 150 years ago as a gold mining town. The town's name is indicative of its location; it is where two forks of the Salmon River meet. The ride to the town from the region adjacent to the Klamath River is an awe-inspiring trip. The road starts as a narrow two-lane road that in five miles turns into a very narrow one-lane road. The road follows the Salmon for its entire trip and also parallels some very steep and high cliffs. The views from the roadway are gorgeous. The Salmon River has a very aqua color to it and is known for its great steelhead fishing in the winter, even though it's only catch and release. In the summer the region is famous for whitewater rafting through its beautiful gorges. As a cautionary tip, do not drive this road if you are sleepy, during bad weather, driving an RV, or you're a little under the weather. It is a dangerous road even on a great day.

Driving this road alone can also be very dangerous. Many of the turns are blind and it's wise to honk your horn and drive slowly. Don't get caught looking at the river as you are driving; pay attention to the road. I stopped more than 10 times to take photos and absorb the beauty and only saw three cars on the excursion. One vehicle was a Siskiyou County road crew.

The mountains surrounding the river are very steep, rugged and, depending on the side of the hill, thick with vegetation or thin and covered with chaparral-type bush. The many times that I stopped, I saw fish in the river and nobody fishing. There was a distinct lack of people in this area; it was a little odd. When I finally arrived in town, the first sign of a municipality was the Siskiyou County Road Department building. I continued to follow the road into "downtown," which consisted of a small gas station, a small trailer that was a store, and a brand new and beautiful United States Post Office.

While I was in town, I made contact with their postmaster, storeowner and some others in the street. I explained to people why I was in the area and asked if they had anyone who had known of Bigfoot. It was at this time that the postmaster's delivery agent told me that his mother, Lillian Bennett, had seen Bigfoot. He explained where she lived and that if I waited a few minutes he would arrange for an introduction.

At about 11:00 a.m. on a cold February afternoon I pulled up the dirt driveway and into the yard of Lillian Bennett. There were a variety of older cars in the yard, which obviously had lived through better days. Lillian exited her house and met me on her porch. She immediately struck me as a woman who appeared much younger than the 89 years that her son had stated. She was quick witted, had a fast pace to her walk and seemed to be a no-nonsense-type person. She escorted me into her home and she took the chair next to a view window looking out into the front of her yard. Immediately next to her seat was a very large and well-maintained 12-gauge shotgun that nobody could miss (the photo I took of Lillian includes the shotgun).

Lillian started the conversation by saying that she was born a Karuk tribe member and she practiced the spiritual ways of her tribe her entire life. She said that she was the mother of 10 kids, five boys and five girls, and was born in a house on this property 89 years ago. She said that she has lived in this region her entire life, except for two years when she lived in Riverside County in Southern California. Lillian said that her dad was a "Jack of all trades," and had to be to live in this area. She said that a majority of the time he would work in local mines, but during hard times he'd do all types of labor to keep his family happy, fed and warm.

Lillian explained that while she was being raised in the area she heard stories about Bigfoot sightings and footprints, and also heard about legends associated with her tribe and the creature. She said that she never really gave it much thought until 1985.

She believed it was either July or August 1985 when her daughter, Shirley Lincoln, was staying with her. Shirley was waiting at her mom's house for another relative who was driving from Eureka to Forks of Salmon by way of the Bald Hills Road. This road was known as being more treacherous than the longer paved route through Willow Creek, but it could also save a lot of time. The Bald Hills route can be confusing because of a lack of road signs. It is probably prudent to never take this route unless you're armed with a slew of United States Geological Survey topographic maps that list all roads and trails in the area. It's also important that you carry food and water in your vehicle in case you need to spend the night out on the road.

Lillian and Shirley were getting increasingly concerned about the absence of their friend and decided they would drive the roads back to Eureka and ensure that their friend's car had not broken down or crashed. Lillian said that very early the next morning they left and started the trip back to the top of Bald Hills. Lillian said that Shirley was driving as they wove their way along the Salmon until it met the Klamath River and then headed west towards Weitchpec. They encountered few vehicles as they passed Weitchpec and continued along to the Bald Hills cutoff at Martins Ferry Bridge. It was sunny as they drove the dirt road to the top of the mountain and waited for a few moments to enjoy the beauty of the summer day; they didn't see any cars on the Bald Hills Road. It was approximately 10:00 a.m. when they decided to drive into Hoopa and check the local motel for their friend and then have breakfast.

The Bald Hills Road turns into pavement as you drive over the crest and head towards Hoopa. Lillian said that she and Shirley had just started their journey downhill when they went around a small turn and saw a Bigfoot walking down the middle of the road in front of them. Lillian said that the creature was about 50 feet in front of their vehicle and was also walking downhill. Lillian said that one memory that will always stay with her is that the creature had hair on its butt that was six to eight inches long and it was swaying as it walked. She said that the creature had the classic Bigfoot appearance, close to seven feet tall with dark hair over its entire body. Lillian said that the creature had very wide shoulders with a narrow waist. She stated that the creature always walked on two feet while on the roadway and its arms swayed a lot as it walked. She said that she has heard stories of Bigfoot sightings and how people relate that Bigfoot's head comes to a point; she said that this Bigfoot's head did not come to a point, but was quite round.

She and Shirley were watching the Bigfoot as she told her daughter to keep the car idling, and not to startle the creature because she wanted to watch it. Lillian said that Shirley accidentally hit the accelerator slightly and that appeared to alert it. Lillian said that the creature never turned completely so that she could see its face. When the creature heard their vehicle, it immediately turned towards the very steep road embankment on the right side and scampered up the hillside until it was out of sight. Lillian told Shirley to drive up alongside the embankment to see if they could see the creature; Shirley refused. At this point Lillian said that Shirley continued the drive directly down into Hoopa where they stopped for breakfast.

Lillian said that when they were in the coffee shop in Hoopa they were talking about their sighting when local Hoopa elders sitting at a nearby table told them to keep their mouths shut. Lillian said that the elders told them to keep their sighting to themselves because they didn't want people coming up here investigating the creature and walking around their forests. Lillian said they were very abrupt and a little intimidating. She and Shirley stopped talking about the sighting at that point.

Near the end of my conversation with Lillian, she gave me the contact information for Shirley and stated that it would be beneficial to the story if I contacted her to verify her account of the incident. Lillian did say that to this day, Shirley would never go up Bald Hills Road again.

Before leaving this section of the sighting I would like to make some comments about Lillian. It was not an easy task to find witnesses in Forks of Salmon. They are a community that is wary of outsiders and their motives. Lillian never told her story to locals, just her family. She never sought me out; I found her through her son. She is a person that struck me as very credible, honest and straightforward. It was a distinct pleasure to meet with her. Lillian did sign an affidavit.

Location of Sighting

The area of Bald Hills Road has had many encounters with Bigfoot over the years, including documented incidents in this book. Bald Hills not only facilitates a roadway from the Eureka coast to the inner mountains, but it also offers a route that animals and Bigfoot can take. It is obvious from Lillian's encounter that Bigfoot sometimes utilizes roads for travel. Bald Hills also borders the region of Bluff Creek and Hoopa, ideally situated for movement by Bigfoot into several key regions where sightings occur. On one side of the mountain sits the Klamath River, and the other has the Trinity River. Each river has large amounts of fish all year long. The forest on Bald Hills is very thick and in places almost impenetrable, ideal for a Bigfoot to live without being seen. There are many similarities between Lillian's sighting on Bald Hills Road and that of Jackie Martins. Both vehicles were traveling downhill and both had a sighting that lasted 5–7 seconds. Both sightings were in close proximity to each other.

The comments from Hoopa elders to Lillian about keeping her mouth shut regarding her story strike me as a factual occurrence. The older members of the Hoopa tribe are much more possessive of their region than younger members. The older members don't want outsiders seeing the resources they own and the land that is theirs. They are probably concerned that outsiders will take what is not their own and cause them loss. The best example of this occurs during mushroom harvesting season. The Hoopa and their tribe members are the only people allowed to harvest mushrooms on their reservation; nobody else can harvest and there isn't even a permit process. Many of the elders have had negative experiences in their lives with non-tribal members and as such aren't as likely to engage an outsider in any type of communication. I've found that if you are honest, straightforward and they see you frequent their reservation on a regular basis, most start to accept your presence and will engage in conversation about almost anything. If you are on the reservation for a day trip or weekend excursion, good luck trying to engage the elders. Once you have their confidence and communication commences, they are great people.

Phil Smith

Owner, Bluff Creek Resort
Orleans

Sighting #11: 1983/1984

Deputy Greg Byers from the Humboldt County Sheriffs had told me to travel to the Bluff Creek Resort and talk to the owner, Phil Smith. The Bluff Creek Resort has been in the Smith family for 46 years. It is located in an optimum location to encounter Bigfoot activity. The resort sits less than three miles from where Bluff Creek empties into the Klamath River and almost exactly at the end of the Fish Lake Road that goes into the Bluff Creek Basin. It is located on the Klamath River on Highway 96 with high, desolate mountains surrounding it on all sides, a typical setting in the Six Rivers National Forest. It is also less than three miles from the confluence of the Trinity and Klamath Rivers, two of the largest rivers in Northern California.

Phil Smith is a 34-year-old former mechanic from Southern California. He was tired of the hustle of city life and took his dad's offer to operate the resort. The property sits directly next to the highway with large trees as a buffer. There are a few cabins, washrooms, showers, some RV parking and even some camping on the lawn is available. In the early days, Phil's grandfather, Lucien Saunders, also operated a large store that serviced the entire area (remnants of the store can still be seen bordering the roadway at the far east end of the resort). The closest stores now are approximately five miles into Weitchpec and 12 miles the other direction in Orleans. In the 1960s, the store was a busy spot for hunting, fishing and camping supplies. Lucien was the person who outfitted the Patterson–Gimlin party for their famous trip when Bigfoot was first caught on film.

When I first interviewed Phil, I asked him to start at the beginning, in the 1960s and walk me through what is happening today with regard to Bigfoot activity. Phil and I sat on his front porch as he enthusiastically told me of his younger years spending weeks with his grandfather at the resort. Phil said that his grandfather was an open book, didn't hold anything back and totally believed that Bigfoot existed. He said that his grandfather thought that Patterson and Gimlin were absolutely nuts to go back into the Bluff Creek Basin with a huge creature running around throwing huge

Phil Smith sitting on a rock wall at his resort holding a photo of a Bigfoot print given to his grandfather by a Humboldt County Sheriff.

construction tires and barrels off cliffs. Lucien felt that it was permissible to casually see Bigfoot, but the thought of a team specifically looking and stalking Bigfoot was a team looking for trouble. Grandpa did eagerly outfit the Patterson–Gimlin team for their excursion and that did give him the small benefit of listening to some of their planning. Grandpa knew that the local Native Americans believed that Bigfoot lived in the Bluff Creek area and it was not to be disturbed. Lucien had told Patterson and Gimlin of the local beliefs, but they didn't seem concerned.

It was amazing to watch Phil tell this story, as he had a special sparkle in his eye and an absolute love of the country and the lore.

When he was approximately 12 years old, the forest service was building the Fish Lake Road, which starts almost in front of their resort. The road is approximately eight miles long and leads to a beautiful trout-filled lake on the edge of the Bluff Creek Basin. The road continues for another 30 miles back into the mountains and eventually makes a huge loop to Orleans, California. The crew had completed approximately 16 miles of the road. There were rumors that the road team had encountered "issues" during the build out. Their construction site was supposedly visited on a regular basis

Photo of picture Smith was holding.

by a huge hairy creature with big feet, coining the name Bigfoot. There were signs of Bigfoot checking out their equipment and even occasionally pulling some shenanigans. Phil said that one weekend his grandfather drove him up to mile 16 in the road build. He said it was still a dirt road under construction and was a rough trip. They left on an early Saturday morning and arrived at the site — lonely, very quiet and a little strange. Phil said that he and his grandfather walked around the construction equipment and immediately observed huge, barefoot tracks wandering throughout the parked machines. He said that the tracks were slightly similar to a person in bare feet, except the footprints were between 15 and 17 inches — huge! Phil said he knew what he was looking at and was a little shaken to see it in person. The location was a little creepy because it was a weekend day, nobody was around and there were tracks everywhere, obviously made since the crews left the night before.

Phil said that a few years went by and he continually visited his grandfather and was updated regularly on the Bigfoot activity. He said that when he was approximately 15 years old on a late summer afternoon, he and his grandfather took a walk behind the resort and onto the bank of the Klamath River with the grandfather explaining the Bigfoot phenomena. He stated that they talked about Bigfoot and the difference between Bigfoot and bears. He said that it is completely normal to see bears at the resort and in the surrounding hills, and that bears have been a constant issue over the years, as they get into the garbage, scare visitors, and essentially wreak havoc. Phil guessed that there have been an average of 10 bears killed at the resort every year. He related that he is very familiar with cinnamon bears and black bears; blacks are usually a little larger.

Phil said that he would tell me a quick story and then get back to his grandfather. One night he awoke to hear something rummaging through the garbage. He walked outside with his 9mm pistol to see the largest black bear he had ever seen in his life. He said he startled the bear, it turned toward him and he shot it once with his pistol, a great shot. He severed its spinal cord and the bear's back legs went limp. He walked closer and completed the kill by putting a round into the back of its head. He was still shaking long after the incident because of the gigantic size of the bear, nearly 600 pounds!

Phil stated that he and his grandfather continued on their walk and made their way into an area near the resort where there is a small creek flowing off the back of the hill on the opposite side of the bank. He said that they rowed their way across the river to the other side and were looking up a shear cliff of rock. Phil said that he and his grandfather were startled to see a huge black figure standing on the cliff above staring down at them. The creature was standing on two legs partially behind a tree. Phil said that this was the first time he ever had seen Bigfoot. He and his grandfather backed off and went home, ecstatic about what they had just seen.

Phil said that probably in the same year as his sighting he was visiting the resort during the winter when he awoke one morning to snow on the ground. He went outside and was shocked to see more Bigfoot tracks in the snow. He recalled that it was probably December or January, very cold and bears were hibernating. He said that the tracks looked the same as the Fish Lake Road tracks, maybe even larger. He could see where Bigfoot would stop and dig briefly through the snow into the grass, obviously looking for something. He went back into the house and his grandfather came out and confirmed his observations. The tracks walked almost right through the middle of the resort area and were still very fresh. Phil stated that those were very exciting times. He awoke every morning waiting to make a new observation or see another Bigfoot.

Smith said that he would still hear Bigfoot screams regularly, and it is almost an expected event at the resort. He said that in late December 2006, he and his girlfriend were awakened at approximately three in the morning by very loud Bigfoot screams that were coming from three different directions and near the tops of three different mountains: Bluff Creek eastern side, Bluff Creek western side, and across the river on the Hopkins Butte side. He said it was obvious that the creatures were trying to triangulate by screaming and thereby find each other. He said that the screams would start, and each would scream for a few seconds and then would stop. They would start screaming again a few minutes later and with each event, they were getting closer and closer to the resort. Phil stated that he and his girlfriend were genuinely scared because of the number of creatures converging on his property. He said that the sounds eventually stopped when it appeared they were all meeting at the intersection of Fish Lake Road and the highway. Phil said that the sounds the creature makes are not similar to any other animal in the woods and cannot be mistaken for a human generated sound. He told me that it wasn't until the sun started to rise that he and his girlfriend could eventually fall asleep.

Phil confirmed that Lucien was a local expert when it came to Bigfoot and tracks. He said that many people would come by the store and talk to his grandfather about sounds, tracks and sightings. At one point a Humboldt County sheriff, Tom Guglaueias, brought Lucien a photo of a track that was left in Orleans. The sheriff wanted to know if it was a genuine Bigfoot track and Lucien confirmed it was (see photo). The track was huge and obviously wasn't made by a human. Phil said that he remembered the day he first saw the photo and it immediately reminded him of the tracks he saw near the road-grading equipment up Fish Lake Road.

Phil was adamant that there are still Bigfoot in the hills surrounding his resort and he still hears them yelling. He stated that you have to spend time in the creek washes, sandbars, basins and valleys looking in areas where tracks have been seen. He also said that you never know when a Bigfoot is going to be staring at you from behind a tree. Phil told me many times that he is willing to take me across the Klamath and show me locations where he believes Bigfoot still lives.

One of the last times I spoke with Phil, he told me that he walked up into the hillside on the opposite side of the Klamath near where he and his grandfather made their sighting. He said that he was walking on a very wide and very high game trail that ran parallel to the river. He said it was an uncomfortable hike and he couldn't quite explain why. He had traveled approximately 200 yards from his start point when he came across a giant pile of scat in the middle of the trail. Phil explained how he had seen hundreds of piles of bear scat in his years in the woods. He stated that this pile was nothing close to bear scat, but similar to a huge human scat pile. He said that the diameter and length of the scat were very large, nothing like a bear or human. He examined the pile briefly, became more uncomfortable, and immediately left the area and went back to the resort. He stated that he would take me back to the spot, but would never go back alone.

Phil did sign an affidavit.

Location of Sighting

Bluff Creek Resort would be considered to be in the heart of the Bigfoot territory. On one side you have the mountains leading to the famous Bluff Creek film of Bigfoot. On another side you have the mountains that form the northern edge of the Hoopa reservation and immediately behind is one of the largest rivers in California, the Klamath. The Bluff Creek Resort would be an excellent location to establish a first class base camp for travels into the surrounding countryside for in-depth Bigfoot research. The resort has just enough amenities to make life comfortable and is close enough to Orleans and Hoopa that you can get food, materials and medical care if needed. This place is high on my list for long-term research.

I had expected to return to the Bluff Creek Resort in the summer of 2007 and take Phil up on his offer to travel across the Klamath to Bigfoot's home. My 2006 summer went by too quickly and Phil and I never got together. In March 2007, I was accumulating my witnesses of Bigfoot and scheduling them to meet with a forensic artist when I returned to the resort to schedule Phil. I pulled into the front of the resort and was met immediately by his girlfriend, Vanessa. I greeted her and asked if Phil was around. She got a very distraught look on her face and said that Phil was dead. My brain didn't register what she said and I again said, "No, I mean Phil the guy who runs the resort, your boyfriend." She again stated that he was dead; he had died 10 days earlier.

I was in absolute shock. She told me that they had a large snowstorm two weeks before and Phil was out in the woods when a fir tree fell over and hit him in the back of the head, killing him instantly. I was numb.

Phil was a man who lived for the woods. He wasn't afraid of bears or snakes — maybe Bigfoot. He was a very warm and charming man whose time on earth was way too short. I owe it to Phil to make that trip to his secret spot. I will do it. Those who frequented the Bluff Creek Resort are better people for knowing Phil, and I feel very blessed for the time we had spent together. Phil Smith and the Bluff Creek Resort will forever go down in Bigfoot history as a centralized spot where nice people lived and Bigfoot frequented. God bless you, Phil.

Jay Jones

Forester
Hoopa

Sightings #12: September 1988
#30: November 2005

Over the years I had always contemplated putting a sign on the side of my 4 x 4 advertising I was with California Bigfoot Search. I had read about other researchers who advertised and had reaped significant benefit. I finally spent the money and got a couple of magnetic signs — and they work!

I was filling up at the only gas station in Hoopa when a young man in a pickup drove up and said that he had seen my truck and sign around town and wanted to talk to me. It was at the end of a very long day of hiking in the backcountry, but I have to make time for the stories.

I completed my fill up and drove around the corner to where my potential witness was waiting. I immediately noticed a huge pit bull dog in the seat next to him. I also saw that he had a chainsaw, fuel and tools in the bed of his truck. The vehicle looked to be over 20 years old and in questionable condition, standard for this part of the country.

Jay Jones introduced himself with a smile. He stated that he was a 34-year-old fire wood salesman and had spent a majority of his life living in the Hoopa valley. He asked me a series of questions about what I was doing in the community and my motives. After the standard series of answers and more questions, Jay appeared to be comfortable and started to tell me stories related to Bigfoot.

Jay said that in his teen years he was quite the troublemaker. He said that there were few things that he and his brother wouldn't try or attempt, legal or illegal. He stated that in September of 1988 he and his brother decided they wanted to go deer hunting with their spotlight up in Redwood Valley, highly illegal in California. Any nighttime deer hunting in the state is illegal. Jay explained that Redwood Valley is just west of Hoopa through the mountains and up Supply Creek, maybe 10 miles. He said that it was approximately midnight when he was in the bed of the truck with the spotlight and his brother was driving with the rifle in his lap. He said that they came upon a large meadow with a few pregnant doe that appeared to be

Jay Jones and his pit bull, Bill.

feeding. He said that they wouldn't shoot the doe and moved the spotlight further down the meadow to see what was in the field. He stated that he lit up a grassy knoll and saw a creature on its knees in the grass. He couldn't tell what it was and he continued to light it up. The creature got up on its two feet and started to run away from him, just like a human. He stated that the creature was 3½-feet tall and dark in color, possibly hairy or covered with fur, but he was not sure. It was very fast, much faster than a man, but it wasn't a man. Jay said that he had no idea what it could possibly be other than a "little person" or a small Bigfoot. They spent a little time looking for the creature but left the area after they couldn't find anything.

Little people in the Indian culture live in the hills surrounding the reservation. They are extremely rare, hardly ever seen and are considered sacred. They do not have feet like people but small feet sometimes covered in hooves. They are normally not over four feet tall, but sometimes can grow to the height of a normal person. They sometimes do have the appearance of a human, but that isn't always the case. I have personally met and interviewed several people who claim to have seen and interacted with the little people. Several months ago I was at Onion Lake near Bluff Creek and met some local woodsmen. One young man (in his 20s) told me he had picked up a hitchhiker out of Orleans several months before. He took the person and dropped him at a location just out of Weitchpec. He said the guy didn't say much and was wearing long pants. When he dropped him off and saw the guy walk away, he saw that his pants had climbed up his leg and he didn't have feet, but hooves. He said that would be the last time he would ever pick up a hitchhiker. There are also stories I have been told by Hoopa residents that say that little people can be mischievous and sometimes dangerous. They will sometimes allow themselves to be seen in the woods so that you will follow them. Stories and legends say that if you follow a little person you may be harmed in some way. Stories about little people are not confined to the region around the Klamath and the Trinity, but are throughout Northern California.

Jay's second incident occurred in November 2005. He was 6–7 miles up Tish Tang Road near the creek. He had been mushroom hunting and searching the forests with his pit bull, Bill. Jay said that he had had been told by tribe members and friends that this was an area that Bigfoot visited, but he really didn't think that much about it. He said he had a small lantern and was looking through the brush. It was a dark night and getting cold. Jay guessed the time was nearing 10:00 p.m. He said that he began to smell a very odd and strong odor on a hillside adjacent to the road. He started to walk in the direction of the smell when Bill turned and ran into the truck. Jay said that Bill is never afraid of anything, bear, mountain lion, raccoon, nothing. He said that it did concern him a little when he saw Bill's reaction to the odor.

Jay explained that he continued towards the odor and got to a small clearing when he saw a five-foot-tall dark mass standing about 150 feet in front of him. He could tell it was dark in color, standing on two feet and not moving. He said it was rock steady on its feet and didn't sway. Jay said he could tell that the creature had a very strong looking upper body with a smaller waist and guessed its weight at 250 pounds; that was about all he could tell. He said that he has spent his entire life in the woods and around animals and what he saw was not a bear — positively not a bear. He guessed he was looking at a young Bigfoot, but wasn't 100 percent sure. He said that once he saw the creature, he slowly backed his way down the hill and to his truck where he left the area. When he got back to the truck his pit bull was laying on his floorboard and appeared to be shaking.

It should be noted that I have met Bill, Jay's pit bull. Bill is a very large, solidly built dog. There is no way I would tangle with him, and the times I was around Bill, he never showed any fear. Bill's reaction to the creature is consistent with many sightings that people have reported with canines and Bigfoot. For some reason dogs are afraid of Bigfoot, and do all they can to avoid contact. Jay stated that he got into his truck and drove back to his house in the valley. Jay signed an affidavit.

Location of Sighting

The location of Jay's sighting is very close to the sighting of Ed Masten and Julianne McCovey. They were all in the woods when their sightings occurred and were not close to roads. They were not in a vehicle, but on foot. I have visited this area many times and have spent considerable time walking the hills and trails. It is very desolate with no residences, but with considerable ground cover, water and even some caves. This road goes all the way to the United States Forest Service Wilderness Area east of the reservation. It is a well-traveled gravel road during daylight hours, but there is almost no traffic during nighttime hours. Ishi Pishi Creek has several access points as you drive up the road, but there are several places where the road is very far from the creek and it would take a significant effort to reach it because of the very steep terrain. There are small trout in the creek.

On November 27 I was driving in the outskirts of Hoopa and saw Jay cutting trees on a hillside. I pulled to the curb and engaged him in conversation. We exchanged greetings, I asked how Bill, his dog, was doing and if there was anything new in the community. Jay said that he didn't know of anything new but did volunteer to take me into the Tish Tang basin and show me exactly where he saw Bigfoot. We agreed to meet on a cold winter morning and make the trek up the valley.

I met Jay early the next morning. It was very cold, had snowed in the hills two days earlier, and there were patches of fog in the valley. Jay stated that I could follow him and he would stop and show me points of interests along the way. I asked Jay to give me an approximate distance from the forks of Tish Tang to where his sighting was made. His answer caught me a little off guard — "45 telephone poles." I got into my Jeep quite amused with his answer and chuckled to myself about the response. Many days earlier I had asked Jay if he could someday show me how his people collect mushrooms in the wild. He said that when we went back into the hills he would show me. Well, we were now en route.

We drove the eastside of the Trinity River from Hoopa until we came upon a fork in the pavement and went east. The pavement soon changed to improved dirt and gravel. After approximately four miles we stopped adjacent to a small hillside that contained scrub oak, large fur and was 90 percent covered in shade with some filtered sunlight. Jay exited his truck with Bill and told me this was his special mushroom spot. He explained that you need some sunlight, not a lot, lots of leaves and good soil. He told me that many people pick tan oak mushrooms to personally eat, while others gather them for Asian buyers. He states that buyers pay as much as $75 per pound for the large tan oaks.

Jay pulled a long pole with a hook at the end from the bed of his truck. Bill looked excited as they both jumped into the hillside and Jay barked orders for Bill to find the mushrooms. Jay explained that you have to look for subtle mounds under the leaves, and this is where the mushrooms are growing. After five minutes he had found his first mushroom, small, but with a very distinctive, unmistakable smell. I won't describe what the mushroom looks like or what it smells like because I am not the expert, and amateurs should not attempt mushroom collecting. After another 20 minutes Jay had 10 mushrooms and enough to make Bill and himself happy. He told me now to follow him to another location where he saw Bigfoot.

We drove six miles up into the Tish Tang basin when Jay again pulled to the side, exited and stated that this is where he saw Bigfoot. He told me that it had been getting dark and he was using a small lantern to look for mushrooms. He explained that the mushrooms are easier to find when it's getting dark because they almost glow under the light of the lantern and dusk sunlight. He said he was half way up the mountain when he saw the figure in front of him. The area we were in had large oak, large fir and an area under the canopy tree line that was easy to walk. I walked into the hillside and found that the area was covered in 3–4 inches of leaves. I shuffled the leaves aside and found a very thin layer of dirt and lots of small rocks, almost like gravel. The area would be impossible to find footprints in because of the soil condition.

Walking in the area where Jay had made his sighting, I was reminded of the area where Ed Masten, Inker McCovey and Julianne McCovey had made their sightings, all while collecting mushrooms. There were definite similarities in the sightings and the circumstances: leaves, trees, grades, etc., were very similar. All three individuals felt that Bigfoot was also in the area looking for tan oak mushrooms. They state that it has been known by tribal members that Bigfoot collects and eats the mushrooms, and this information has been passed down several generations. In two of the three sightings the people were collecting alone. All three stated that there were no other people anywhere in the area when they were collecting. All of the locations have the same distinctive forest appearance. All of the locations have oak trees, moss on the ground, some fir trees, lots of shade and filtered sunlight and the hillside is facing southeast. The elevation of this sighting is 2000 feet.

Jay Jones signed an affidavit.

Joe O'Rourke

Eureka

Sighting #13: Summer 1992

Joe O'Rourke is a Yurok tribal member who spent his early youth practicing the customs of his tribe. Joe was originally referred by members of the Hoopa Tribal Police as a witness on the Raven Ullibarri Bigfoot sighting. Joe was one of the first Hoopa tribal police officers on the scene. Raven had also referred me because Joe was at the scene of her incident. Tribal members also told me that Joe had a Bigfoot sighting when he was a young man and he was very credible source.

Joe had moved from the Hoopa area and quit the police department because of low pay. The Tribal Police in Hoopa are very professional and outstanding individuals, but they are extremely underpaid for the region. Almost every other county and municipality police in the region make more than Hoopa police. Police officers work under the county sheriff and are deputized as Humboldt County Sheriffs working in a Hoopa Tribal Police uniform.

It took me some time, but I eventually found Joe working in Eureka and was able to set a meeting to discuss the multiple incidents.

Joe is one of those people you meet who will impress you with his straightforward approach to answering questions. He is very polite and articulate. He looks and talks like a cop, but he also has a side to him that would remind you of the kind-natured approach to life that you see from the tribal members in the mountains.

I initially talked to Joe about the Raven Ullibarri incident and if he could remember that night. Joe stated that he would never forget the incident and how it went down. He remembers that he was responding to a trespassing or prowler call at Upper Mill Creek Road, where Raven was the reporting person. He knew the house well because of the ties that Raven had to the department. He stated that Raven lived in the last house on the street before you go up and into the mountains. Joe said that he knew that his sergeant, Joe Masten, was also responding to the call and actually was the first on the scene.

Joe O'Rourke, Eureka.

Joe can remember arriving, exiting his police cruiser, and feeling the strong vibration from a huge creature walking and leaving the area. He stated he went immediately to the porch where Raven and his sergeant were talking. The conversation was brief. Raven pointed where the creature had escaped, and then he and the sergeant decided to follow the creature's path into the forest to see if they could find it. Joe stated that he had his pistol out as they entered the forest canopy at the edge of the Ullibarri fence line. Joe made it clear that as they were walking towards the forest, with every step they took they could hear the huge thump of the creature taking a step, almost as though it was trying to match their steps. Once he and the sergeant entered the forest, it seemed that all sound and noise ceased and only they were making the noise. He stated it was very, very odd, almost like a vacuum had taken everything out of the air.

They searched for approximately 30 minutes and never found any evidence of the creature. Joe said that he never smelled anything odd and he didn't remember the wind blowing. Joe told me that as he was searching, he always felt as though something or someone was watching him.

In one of three different conversations I had with Joe, I specifically asked him if he had ever looked up into the canopy to see if there was something hiding in the trees. Joe stated that he thought about looking up, but he said, "You know I had thought about that, but sometimes you just know you don't want to look up."

After searching the area with the sergeant, they both returned and talked to Raven more about the incident. Joe stated that he heard the description that Raven had given them — huge creature, walking on two legs, hair on almost its entire body, huge shoulders, walking on two feet, carrying garbage bags in its hands, etc. He knew that Raven had just had a Bigfoot in her backyard. Joe specifically stated that he felt that this creature was very intelligent. He explained that there were radios blaring, bright lights from their cars and officers in uniform, yet the creature remained calm and was able to elude them. He stated that he has grown up around animals that are native to this area, and every other animal he knows of would have fled the area immediately and not tried to hide, as he believes that Bigfoot did on this occasion. Joe said that there was no way the Bigfoot could have escaped from the immediate area of Raven's residence in the short time it had by itself; it had to have been in the immediate area.

Joe said that he remembered talking to his sergeant after the incident and the sergeant telling him that it probably was a Bigfoot and, based on Raven's description and what they felt and heard, it was probably in excess of 700 pounds. The sergeant also told him that the neighborhood had a history of Bigfoot sightings and the Raven incident didn't surprise him.

Joe's own Bigfoot sighting occurred when he was 12 years old, in 1992. He was one of many youngsters who were attending an annual Yurok Jump Dance, a tribal summer custom. The dances incorporate many of the traditions of the tribe and try to bring the youngsters into the realm of the ceremony. The dance is held near the banks of the Klamath River in Pecwan, west of Weitchpec. The dances are always held during the summer when the temperature at night is warm and the days are long. Joe specifically remembered that he was with a group of family members and friends, and they were at a break in the ceremonies. He stated that he and his cousin, Mia Inong, were walking near the banks of the Klamath on the northern shore. Joe said that he believes that Mia was 10 at the time and they might have been with others, but he wasn't sure. They heard something walking on the other bank, turned and looked, and saw a huge creature walking on two feet. The creature had huge shoulders, hair or fur on its entire body and was walking more like a human than anything else. He said that they never saw it on anything other than its two feet. They only saw it for 3–4 seconds before it disappeared into the forest line.

Joe stated that he has seen many movies and videos that depict Bigfoot, and that the creature that they saw matched the exact description of the Bigfoot shown in the videos and movies. He did emphasize that it was dark outside and there was possibly a full moon, but he couldn't offer any other specific details because they were almost 300 feet from the creature. Joe was absolutely positive the creature wasn't a bear and was a Bigfoot.

Joe signed an affidavit on the details of his sighting on the Klamath and on the details of the Raven Ullibarri incident.

For full details on the Raven Ullibarri incident go to sighting #21.

Juliene McCovey

Hoopa

Sighting #15

Juliene knocked on my hotel room door one day and said that she had heard that I was a trusted investigator/researcher and she wanted to tell me her Bigfoot story. I asked her to wait by my Jeep and that I'd be with her in a few minutes.

I interviewed Juliene in the parking lot of my hotel, adjacent to Ray's Supermarket. (Bigfoot research isn't very glamorous.) Juliene stated that she had spent the majority of her life in Hoopa, but also spent many years in Oregon going to school. The job she had when she made her Bigfoot sighting was in the fire fighting division for Hoopa Forestry.

The day that Juliene saw the Bigfoot she had been dropped in the corner of the reservation in an area called "Horse Linto." She said that she and her partner were to walk the reservation in a specific angle and then meet the remainder of the crew on the Tish Tang Road. They were dropped at 6:30 a.m. and they started their walk north.

Juliene recounted that she and her partner were above Tish Tang Falls and attempting to cross a very steep and hazardous area. She explained that her male partner crossed first and then was reaching out to pull her across. As she reached out, he grabbed her and pulled her across, but he then lost his balance and slid down into the canyon, completely out of sight. It was just steep enough that he couldn't stop, but slight enough that it wouldn't injure him as he was sliding. As she couldn't get down the hillside to assist him, Juliene started to follow a game trail towards the area where she was to meet her team and could get some help. She saw some heavy cover approximately 60 feet in front of her in an area that caught her attention because of some shade. Under the cover and sitting on the ground she saw a large Bigfoot that appeared to be eating a plant.

Juliene explained that the creature was sitting on a slight hillside with the feet facing downhill. She said that it appeared to be eating a fern. The creature would grab a piece of the plant and then drag it through its mouth in an apparent attempt to eat just the leaves, or taking the moisture off the leaves, Julienne wasn't sure. She also stated it was difficult to judge its height because it was sitting on the ground. She also was very sure that the creature was either pregnant or had something laying or sitting on its lap because the belly of the creature was huge and seemed disproportional to the rest of the body. She described the body as thick with large shoulders and black hair. The other unusual element to Juliene's description is that the creature had 10-inch long hair streaming from its head — 10 inches!

Juliene explained that she walked at a slight angle towards the creature reaching a point as close as 20 feet. She said that the creature never appeared to be frightened of her, and never moved from its position. Juliene said that she said "Hi" in Hoopa, and then started to walk up the hillside and away from the creature. When she spoke, it appeared to her that the creature smiled. Juliene also wanted to make sure that I understood that what she saw was a female Bigfoot. She told me that she feels sure that it was a female based on its soft facial features, and the manner it held itself, and the polite and sweet attitude it expressed. It was distinctive.

Juliene McCovey at the Tsewenaldin Inn, Hoopa.

Approximately 10 minutes after seeing the creature, Juliene met up with her fire crew. She said that she told them what she had seen and that she was sure it was a Bigfoot. Juliene said that nobody on the crew believed she had seen the legendary creature.

Juliene signed an affidavit.

Location of Sighting

The sighting took place in the Tish Tang basin. This is a place that has had a history of sightings for many years. Ed Masten and Jay Jones had multiple sightings in this region, and many aspects of the sightings fit the profile. It's obvious that this creature was not afraid of Juliene and made no effort to leave the area, an unusual behavior for Bigfoot. Maybe it was the female–female issue, where they both knew they shouldn't be afraid of

Juliene's sketch of the creature she saw in the Tish Tang region.

each other. In questioning Juliene at length about the creature's activity, it appears that it was eating some type of fern by pulling it off the main plant and then dragging the plant through its lips. Julienne made it clear that the creature was sitting in a shady area that had many ferns adjacent to where it was eating.

Forensic Sketch

Juliene came to the hotel and met with Harvey and Gina. She was very sure of herself on the description and took little time putting together the basic facial structure that Harvey drew. The drawing does have a soft feel to the facial area and does appear to be different than other drawings that other witnesses have felt to be male Bigfoot. Juliene stated that the creature had some hair on the face, but little, and long flowing hair from the head.

Juliene took a significant time explaining how she saw the creature sitting and taking the branch or fern and then dragging the leaves through its mouth. She said that she isn't 100 percent positive if the creature ate all of the leaves, or was just taking the condensation off the leaves and utilizing it as water. Once when I was watching the Discovery Channel, they showed footage of an ape in the wild. The ape was dragging leaves through its mouth in the same exact manner that Juliene was describing Bigfoot dragging the fern through its mouth.

When Harvey had finished the drawing, he showed it to Juliene and asked for final comments. Juliene told him that it was a great rendition of what she saw in the woods. It's an interesting drawing because it possesses female characteristics and there is softness in the sketch. I think Harvey and Juliene worked well together and Harvey was able to add a feminine touch to the huge creature.

Ed Masten

Recreation Group Manager
Hoopa

Sightings #15: November 1978
#30: November 2005

On October 16, 2006, I made the eight-hour drive to Hoopa from the Bay Area. The weather got progressively worse as I got closer, and my hopes of doing fieldwork faded quickly. I arrived at the hotel, checked in and found that there were no new sightings in the last week that had been reported to my contacts. It was now pouring rain and I decided to contact Inker McCovey at Tribal Recreation.

I made the short drive to the recreation center and found Inker talking with some adults about the upcoming basketball league. He was happy to see me and I was immediately introduced to one of his workers,

Ed Masten. The last name immediately struck me as a name I knew. A cousin of Ed's was Sgt. Masten who responded to the Bigfoot rummaging through the garbage at Raven's house off Upper Mill Creek Road.

Ed is a 48-year-old Hoopa tribal member who has spent his entire life on the reservation. He had weathered skin, but was in fantastic shape. Inker introduced me as the "Bigfoot Man" and said that Ed had some stories for me.

Ed started the conversation by stating that he'd had a few Bigfoot encounters and he's not sure why he has been so fortunate. He did preface by stating that he spent a lot of time in the woods, hunting, collecting mushrooms, fishing and netting. He said that he enjoys the outdoors and appreciates what it has to offer. He stated that he was married, has a daughter, and his wife is a registered trauma nurse in Sacramento.

Ed stated that his family has the rights to the fishing grounds adjacent to the Tish Tang Creek and Campgrounds at the Trinity River. He said that he has spent the majority of his life visiting the area and has seen every type of wildlife in the region that anyone could imagine. Ed has seen 600-pound black bears and has even seen an albino (white) black bear that had to weigh over 500 pounds (Elaine Creel, Hoopa Wildlife confirmed that they exist). He said that he even shot and killed a black deer that is now used in ceremonial dances. Ed said that his life has been filled with much luck and enjoyment.

I should state at this point that while Ed was talking about the Tish Tang area, my memory was rocked when I remembered past Bigfoot reports of sightings in the area. Every time I drive into Hoopa from Willow Creek, I drive past a public campsite called Tish Tang Campgrounds, and Bigfoot tracks had been reported here many years earlier. It's an unusual name and the sighting stuck with me. When Ed started to talk about multiple sightings in this area, it immediately struck me as a possible credible sighting.

Ed's first Bigfoot encounter occurred in 1978. He remembers it occurring in November, but doesn't recall the exact date. He was walking with his brother near the road that runs adjacent to Tish Tang Creek and the Trinity. He said it was later in the afternoon, between 3:00 and 4:00 p.m. He smelled a slight odor and started to slow his walking pace. The odor started to get worse when he saw a creature partially hiding behind a tree 15 feet away. He said that the creature was standing on two feet and

Ed Masten outside the recreation office in Hoopa.

peering around the tree at them. The creature was huge, almost 10 feet tall. It had hair over its entire body, and had huge arms that wrapped almost completely around a large diameter tree. Ed said that the creature had a face like a Bigfoot, not a bear. The creature did not have a big snout, never walked on four legs and had eyes that were piercing. Ed stated that he can distinctly remember that the creature had eyes that were a deep orange in color. He said that he will never forget those eyes. Ed said that he was within 15 feet of the creature when he first saw it; he looked at it for a few seconds and then started to back away. He and his brother left the area that night after the observation.

Ed's second encounter again revolves around the Tish Tang area. Ed stated that he and his brothers and father have netted and fished this area for generations. He volunteered to take me to the area. When we arrived I found it a very remote location, at the end of one branch of a dirt road that runs along the eastern bank of the Trinity River. Many of the tribal members' fishing grounds are located in the middle of the community with little or no commute to the site. This location is beautiful, isolated, and the adjacent forest is very dense. When we arrived at the site Ed wanted to show me a rock that was sitting partially in the water. He stated that the rock appears to have a bone, skin and possibly other bones buried in it. He showed me the rock and it did appear to have a joint with bone that was somehow inside the rock. It was an odd site and a strange find. It was located very near the location where the Tish Tang Creek empties into the Trinity River. It was one of those very odd occurrences that happen in this valley. Prior to delivering the rock to Hoopa geology, I had the specimen examined by a geology professor from the University of British Columbia. He stated that the rock came from a cave and probably washed down the creek. This rock was later given to the geologist who works for the tribe.

Ed said that his second encounter occurred in November 2005. He said that he was up in the hills directly above our location on the Trinity. He was with a friend, Vickie Grant, and they were mushroom hunting. He said that he has made it a point in his life to be observant of wildlife and their tracks when in the wild. He stated that this day in November was no different. They were up a hillside where he could see that a huge animal had come down the embankment and left a large pile of compressed dirt near the base. He could tell it was not a bear because of its size, amount of compressed dirt and the number of tracks that were left. They continued walking and came upon a small area of land that was wet and slightly mossy. He said that he saw several huge footprints, similar to human, but gigantic. The toes were much more round than a human's and the depression that the print made was much deeper than he and his friend could make. Ed said that the stride was approximately seven feet. He stated that the prints were 16–18 inches long and much wider than a human print. In fact, the print was so large that his boot could fit entirely inside without touching the perimeter.

While they were examining the footprints, a very pungent odor that was getting stronger by the minute, blew into the area. Ed said that he has smelled human bodies that were left in the river for days and that odor isn't as bad as the odor he and Vickie were experiencing. He said that it was something like ammonia but worse, much worse. Ed said that he was positive that there was a Bigfoot somewhere close by. He and Vickie decided to leave the area when the odor became overwhelming.

Ed explained that he has seen indications over the last 30 years that Bigfoot is using the Tish Tang area as a pathway to and from living areas, food gathering points and river access. He stated that he has seen large tree limbs twisted off and broken where there is no other animal that could do it. He said that there were several occasions where he had been down to the river in the Tish Tang region and he felt as though he was being watched.

After approximately 20 minutes examining the rock and possible bone, walking throughout the area and looking at the creek, Ed and I left the area and I transported him back to the recreation center. Ed said that he'd be happy to escort me up into Tish Tang Creek and show me the locations where he had seen prints and bones in the past, and even a cave.

Two days later Ed guided me into the Tish Tang drainage area on the eastern slopes of Hoopa. It is a very rich environment with a year-round creek that feeds the valley. Salmon and steelhead commute up the river to spawn, and local bears and deer find these mountains a great place to live. Ed told me we would need to travel 8–9 miles up into the backwoods to get to an area where he had been searching for mushrooms. We spent approximately 45 minutes making our way through areas that appeared to have had no vehicular traffic in days. Once we arrived in a dark corner of a valley, Ed directed us up the hill into a very heavily forested area. There wasn't a lot of sunlight in this region and it appeared to be a great place to collect mushrooms. Ed explained how he had been looking for mushrooms, found the prints and then the smell. There were marshy areas and mossy patches and even spots where mushrooms grew. We searched the area for several hours and could not find any evidence of Bigfoot. Ed signed an affidavit.

Location of Sighting

This area of Tish Tang has miles of open and empty forests to its east, and the Trinity Alps Wilderness Area also borders to the east. We could have traveled another 2–3 miles before the dirt road stops and the wilderness area begins. There are no vehicles allowed in the wilderness area, and the United States Forest Service manages it. This would be an excellent area

Ed Masten's sketch of the creature, Ed was very pleased with Harvey's work and felt this was a great representation of what he observed.

for Bigfoot to live without its dens being bothered by humans. There are few tribal members that venture into the wilderness area because it is very rough travel. Members of the public cannot access the wilderness from this side because they can't travel through the reservation. It would be a safe bet that few people are walking in and out of the wilderness area by way of the Tish Tang Creek Region.

On our drive back down the mountain, Ed asked me to pull to the side of the road near the Tish Tang creek. He pulled out fishing line and tied a homemade fly to the end. He said that he wanted to show me that trout were in the creek. Three casts (throws) later, Ed had a beautiful seven-inch trout that he released back into the crystal clear waters of Tish Tang. It was a great day with a very polite man.

The area of Tish Tang is also near the area where several incidents of large boulders being thrown into the Trinity River were reported (refer to Corky Van Pelt and Hank Masten statement). Julianne McCovey also had a sighting in the same valley approximately two miles from Ed's sighting.

Forensic Sketch

Ed was very eager to help us with this project. He made time in his busy schedule with Hoopa Recreation to assist Harvey with the preparation of the sketch. Ed started by stating that he felt the creature was 9–10 feet tall and weighed 500–600 pounds. He described the eyes as being a shade of orange, and the creature having short hair on its face. He said that the creature had hair on its entire body and said it was a bluish-brown in color. He stated that there was a strong urine or ammonia odor in the air. He also described the creature as having forearm hair that was 4–6 inches in length and was shiny. He said that it appeared to him as though the creature had no neck, and he felt it was probably a male even though he saw no sex organs. He also stated that as the creature ran from him, he saw that there was a discoloration on its butt as though it had been sitting on the ground.

Page Matilton

Teacher
Hoopa

Sighting #16: August 2004

Sometimes in life people find you for reasons that aren't apparent at the beginning, but flower into reality later. Page Matilton initially contacted me at my hotel and introduced himself as the sixth grade teacher in Hoopa. He stated that he had always had an interest in Bigfoot and had listened to the elders' stories and tried to live by tribal beliefs. Page asked me if I would take the time one day and make a presentation to his class about the type of work I was doing in the community and the type of information that I had developed. I told him that I would enjoy doing that and we set a date.

Page's class was extraordinary. The kids asked great questions, they were interested and they even had some of their own stories and beliefs. Page has a deep interest in Bigfoot, and from the beginning I knew that he was a believer. Sometimes I think a grammar school class starts to take on the personality of their teacher, and I think this class did with Page. He is outgoing, very polite and quite inquisitive, just like the members of his class.

After I gave the presentation to the class we had an open segment where everyone asked questions and told their personal stories. Page told his.

Page stated that he had grown up in the valley and spent countless days in the woods surrounding Hoopa, and was always looking for signs of Bigfoot. He stated that he knew that the elders felt that Bigfoot was a friend, would not harm them, and took care of the woods. Page said that all tribal members were meant to believe that Bigfoot was out there, but you'd really never see him.

Page Matilton behind his desk in his sixth grade class in Hoopa.

Mr. Matilton said that one weekend his family wanted to go out for a drive. They were driving on Highway 299 outside of Willow Creek near Salyer and they had crossed over at least two overpasses after leaving Willow Creek. He stated that he thought it was about noon when they drove over another bridge. He said that for some reason he looked to his right (he can't explain why) and observed a larger, dark upright figure leading a shorter dark brown figure by about two strides. Page's son simultaneously stated that he had just seen Bigfoot down by the river. There were vehicles behind them, so they couldn't stop on the bridge.

Page said that he talked to his son and he stated it appeared that the larger Bigfoot was a head taller than the short one. Page said that they all tried to talk out the story after they had passed the creatures. They all agreed that bears do not walk upright like they had just seen, and people are not completely covered in dark hair or fur. Page also stated that the two creatures they saw were much too large to be human beings.

Page signed an affidavit.

I have been back to Page's class on two separate occasions. I have also seen many of his students throughout Hoopa and they always come up, introduce themselves and ask great questions. Page really has the pulse of the tribe and knows many of the issues in the surrounding area. I have continually kept in contact with him, as he is a good person and a great teacher!

I know that I have gotten more out of visiting with Page's class than they ever got out of me. Thanks, Page and kids.

Clifford Marshall

College Student
Humboldt State University

Sighting #18: December 2003 or 2004

The Hoopa Valley Indian Reservation is private and not open to the general public without approval from the tribe. The public is allowed to travel on the paved streets without restriction, but cannot enter their dirt roads without specific written approval. As part of the approval process, I went to the tribal offices and met with their chairman, Lyle Marshall. He is a very polite and intelligent person who actually had some level of interest in Bigfoot. I had briefly described my rationale for wanting to visit their forests when he interrupted and stated that I might want to talk to his son, Clifford, because he claimed to have seen Bigfoot. I thought to myself that this was fantastic, the chairman's son claims to have seen Bigfoot; wow, the chairman has to be a believer. The chairman ordered his staff to issue my permit and I asked him how I could contact Clifford. He explained that Clifford was a student at Humboldt State University and gave me his cell phone number.

Clifford is a 21-year-old student who graduated from Hoopa High School. He has a younger sister who also lives near him on the coast. His family lives in Orleans. Clifford is a Yurok tribal member who grew up living in Hoopa.

A few days after meeting the chairman, I called Clifford by phone and we spoke about his sighting. He stated that it was either 2003 or 2004 in December when he was driving from Weitchpec to Hoopa with a friend,

Clifford Marshall at the "Plaza" in Arcata.

Jessica Allen. He said that it was approximately 9:30 p.m. and they were traveling south on Highway 96 near the bluffs just north of Hoopa. They had traveled 80 percent of the distance to Hoopa and were in an area where rock walls come straight down onto the roadway and a sheer cliff with a barricade is on the other side of the road. This portion of the roadway is very treacherous, curvy, and a place where it would be easy to get hit by falling rock or even to drive off the cliff if the barricade didn't stop you.

Clifford said that Jessica was a front seat passenger and they were traveling very slowly. He specifically remembered that the vehicle's high beams were on. He said that they were in a very curvy section of the bluffs when they came around a sharp corner and both saw a creature kneeling against the rock wall in the northbound lane of traffic. Clifford said that initially he thought it was a bear because it was in a hunched position and it was dark in color. He stated that his perception changed very quickly. As the lights of their vehicle illuminated the creature, Clifford saw it stand straight up on two legs and start to walk quickly at an angle southbound

Cliff where Bigfoot jumped.

and towards the cliff in his lane. Clifford said that he slowed his vehicle as he and Jessica watched the creature move towards the railing.

Clifford stopped his story at this point and went into a detailed description of the creature. He stated that it had the classic Bigfoot appearance, as tall as six feet, five inches in height, had hair over its entire body, very wide shoulders and a big head. Clifford said that it appeared that its arms hung lower down the leg than human arms, and the creature moved differently than a human, but nothing like a normal animal or even a bear. Clifford made it very clear that growing up in Orleans and Hoopa, he has seen many bears, and the creature he saw was not a bear.

Clifford said that he and Jessica watched as the creature made its way to the barricade at the far right portion of his lane. The creature put its right hand on the barricade and threw its legs over and simultaneously went over the cliff and disappeared. Clifford said that he couldn't believe what he was watching. He stated that it was at least 200 feet straight down to the Trinity River in this area, and that no man could survive the fall.

They continued driving southbound into Hoopa and didn't see another vehicle until they came near the city limits. Once back in the city, Clifford went to a friend's house and got three of his male friends. They went straight back to the location where Clifford saw the creature jump. He states that they brought high-powered flashlights and couldn't see anything lying at the bottom of the cliff. Clifford did confirm that the distance down to the river was close to what he guessed, 200 feet.

I asked Clifford what he believed he and Jessica saw that night on Highway 96. Clifford said that he is 100 percent positive that they saw Bigfoot. He stated that the way the creature moved, its hairy body, size and ability to survive a leap of the magnitude that it took, it had to be Bigfoot. Clifford went on to explain that there comes a level of responsibility and honor in being the son of the Hoopa Tribal Chairman and that he would never stretch the truth about an issue such as Bigfoot. Clifford signed an affidavit.

Location of Sighting

I've driven the area of this sighting many times, and it is very hazardous. The cliffs above the roadway are several hundred feet high and the cliffs leading into the Trinity River fall several hundred feet to the bottom. There are areas on either side of the cliffs where a creature could access the roadway. There are small creek beds on each side that lead down to the road and these are access points from the Upper Mill Creek area. This sighting location makes a lot of sense when you look at where it is in relationship to other sightings. It is directly across from Transmitter and Bald Hills and the location of many sightings and incidents, and it is just below Upper Mill Creek and a multitude of other sightings. The real question that I've always asked myself is, did the Bigfoot know it was jumping several hundred feet off a cliff when it made that leap? If Bigfoot is a creature that is eight feet tall and weighs 800 pounds, then its bone density must be much greater than a human's. It is still hard to believe that the creature could survive a fall of 200 feet unscathed.

The photo shows the area where the Bigfoot made its leap with the Trinity River at the bottom of the gorge. The background shows the gentle slopes of the mountain where a Bigfoot could easily get to the roadway. Clifford and I talked at length about why the Bigfoot would put itself in the position it did where it really had no area for a safe escape. The only answer we could both come up with is that it might have been collecting road kill.

Mary McClelland

Karuk Member

Sadie McCovey

Yurok Member

Sightings #19: November 2003
#24: Spring 2004

Mary is an attractive 27-year-old mother of one son and now lives in Somes Bar, California. She is a Karuk tribal member who was raised in Willow Creek and attended Hoopa High School.

Inker McCovey referred Mary to me. Inker stated that Mary was with him when they both saw Bigfoot on the highway just outside Hoopa. He gave me her contact information and urged me to call her.

I was able to meet and interview Mary in October 2006 in Orleans, California.

In November 2003 Mary was dating Inker McCovey. She was a passenger in Inker's vehicle when they were headed to the Hoopa Casino at approximately 11:30 p.m. She said that they were traveling westbound on Lower Mill Creek Road and came to the intersection with Highway 96. She said that there was no traffic anywhere and they made their turn to go southbound towards the casino. She said that as they were starting to accelerate over the Mill Creek Bridge she saw something in the roadway about 75 yards in front of the vehicle. She stated that it appeared to her that the creature was bending down on one knee to grab something with its hand. As the headlights of the car illuminated the creature, it stood up, turned slightly and ran eastbound off the roadway. She stated that the strides were huge.

I questioned Mary about the position that the creature took while it

Mary McClelland. Sadie McCovey.

was on the ground and asked her to explain further. The creature was bent over with one knee closer to the ground and it appeared to be picking up road kill. She said that it was facing west across the roadway in the middle of both lanes. She explained that when the headlamps of the car fully illuminated the creature, it turned away from the lights, made a 180-degree turn and ran off the road in an easterly direction. I again questioned Mary regarding the "about face" it made in the road, not normal behavior for an animal as it would normally flee in the direction it was facing. Mary was adamant that the creature made a full 180-degree turn and ran in the opposite direction it was facing. She said that it was amazing how fast the creature got off the road and into a safe area.

Mary said that the creature had hair covering all parts of its body, with possibly just short hair on its face. According to Mary, the hair on its knees and calves was longer than hair in other areas by maybe 3–4 inches. She said that she would guess the creature was 7–8 feet tall; its arms were longer than human arms and the creature's shoulders were wider than its waist. The creature was not heavy and appeared rather lean. Mary said that she would not say that the creature ran away, but that it moved quickly and jumped a small fence into a friend's backyard that fronts Highway 96. Mary confirmed that the creature had a human-type face, not animal. It did not have a snout like a bear and only moved on two feet, much like a human would stride or walk.

Mary said that they continued to drive for several seconds and then simultaneously stated, "Did you just see that?" They both agreed that they had just seen Bigfoot. Mary said that the build, size, dexterity and ability to move were nothing like a bear, they were much more human-like. She also stated that she is 100 percent convinced that she and Inker had just seen Bigfoot. Just as I was finishing my questioning and thanking her for her time, she sheepishly said there was another event. It should be noted that it is not unusual for witnesses to only explain a singular incident that an investigator is questioning them about. If the investigator does not ask if there were other incidents, the chances are that other sightings would go unstated. Thanks to Mary and her comfort level of talking about sightings, she volunteered her information.

Mary was a little shy about talking about her sighting three months after her first. I told her that all sightings were important and her honesty was appreciated.

McClelland stated that it was in the early spring of 2004 when she and Sadie McCovey were driving from Hoopa to Eureka. Mary said that her boyfriend at the time (Inker McCovey) was coaching the boy's basketball team at Hoopa High School and they had a game later in the afternoon in McKinleyville. She said they wanted to ensure they had enough time to get to the game, so they left early in the afternoon. They drove from Hoopa to Willow Creek and were driving the Berry Summit route down to Mckinleyville. It was a sunny day with little traffic. About 1:30 p.m they passed the vista point lookout (Berry Summit). They were approximately three turns further downhill when they noticed a deer running on a bluff above the roadway. Mary said that the deer caught her attention, but what was chasing it made her pull her car to the side of the road. Mary said that she was having another Bigfoot sighting and this time she was with Sadie. She said they saw the Bigfoot chasing the deer near the top of the Bluff and that the Bigfoot was 25 feet behind the deer when they initially saw it. She guessed that the chase was happening 600 feet from their position. They were both stunned as they watched the Bigfoot close the gap on the deer and chase it around a tree, yes, closing the distance between itself and the deer — remarkable. Mary guessed the deer was a yearling and the Bigfoot was close to seven feet tall. She said that she was shocked how fast the Bigfoot could move and not just keep up with the deer, but also make up ground between them. Mary said that this Bigfoot looked a lot like her first Bigfoot sighting, except this one had dark brown hair/fur.

McClelland stated that they watched the chase for close to 20 seconds when both creatures eventually disappeared over a ridgeline. Mary said that she and Sadie looked at each other and stated, "Bears can't run on two feet like that." Both agreed that they had just seen a Bigfoot trying to catch a deer. They pulled back onto the roadway and continued their journey into Eureka.

Since her sighting on Berry Summit Mary said that she heard of others who have seen Bigfoot cross the roadway in that area, but didn't know their names. She explained that the Bigfoot chasing the deer ran similar to a human, but the way its arm swung and the speed on its feet were quite different than any human. She said that the deer was in a life-or-death race to get away from the Bigfoot and this should explain how fast the Bigfoot was moving — very fast.

Mary gave me the contact information for Sadie and I was eventually able to interview her. Sadie is a very intelligent young woman who knew exactly the incident I wanted to talk to her about. She stated that she and Mary had definitely seen Bigfoot and nobody could convince her of anything else. She stated that they had just crossed over Berry Summit and saw movement to the right of their car. She turned and she saw a deer being chased by the Bigfoot through a meadow. Sadie said that she knows that it sounds strange, but it's true and the Bigfoot was almost making up ground on the deer. Sadie said that she has never seen anything like it before in her life and was shocked to have seen it. She said she has seen 100 bears in the wild and is 100 percent certain that it wasn't a bear, and it was much too large to be a human. Sadie said she is positive she saw Bigfoot.

Sadie and Mary each signed affidavits on the Berry Summit incident and Mary signed an additional affidavit on the sighting on Highway 96 near Mill Creek Road.

Location of Sightings

The location of Mary's sighting on Berry Summit is a place I know very well. I cross this summit each time I travel to Hoopa from my residence in the Bay Area. The road from Eureka to Willow Creek is a winding, predominantly two-lane road that starts at sea level and reaches almost 3,000 feet. Early in my travels I noted that this road had few residences on it, and had a few other unusual characteristics. I saw many deer, squirrels and other wildlife on my drives in other parts of California. I also saw blood splatter on the roadway where cars and trucks had struck wildlife. I never saw a dead animal on the road, shoulder of the road or in turnouts in this area. Could the reason for this be that Bigfoot gets its meals from monitoring the roadway and grabbing road kill?

The trip from Eureka to Willow Creek is beautiful, with great views and many mountain vistas and meadows that are gorgeous. Once you've made this trip a few times you can believe the story Mary recounted as it matches the hillsides and landscape perfectly. There are several huge meadows in this area, as well as several large rivers and creeks that could easily support large mammals.

The location of Mary and Inker's sighting on the highway just outside of Hoopa is a heavily traveled route between Hoopa and Weitchpec and Hoopa and Orleans. This location is also part of the group of sightings that fits into a small geographic region that runs from Mill Creek Road, across the Trinity River and up to Bald Hills Mountain that I coined "Bigfoot Alley." The number of sightings in this area is truly staggering for an urban setting.

Forensic Sketch

Mary was always enthusiastic about working on the sketch of the creature. Once the time came to meet, Mary had family time issues and initially stated that she couldn't get to Hoopa from Orleans to meet with us. I offered to bring Harvey to her to save time. She accepted the offer.

We met in the backroom of the only restaurant and bar in Orleans. We sat in a well-lit area in the back where we had solitude and good lighting. Mary was a little concerned that she and Inker McCovey saw the creature and that we may have each of them do a drawing. She told us that she had a much better memory than Inker and that we should count on her. We all had a good laugh.

Mary has always been very sure about what she saw on Highway 96 at Mill Creek in Hoopa. She told us that she felt the creature was near seven feet tall, had huge, muscular legs and buttocks and had grey near its shoulders, with the remainder of the body being brown. She stated that it appeared to her that the creature was wet, as though it had just left the river or creek. She described the arms as much longer than a human's, hanging down to near its knees.

Mary said that when they initially saw the creature it was kneeling on the ground facing the Trinity River, west. She said that when their lights illuminated the creature it made a 180-degree turn and ran east towards a house on the east side of the road. She said they watched the creature jump the fence into the neighbor's yard and disappear. She noticed that it had something in its hand that might have come off the road. Mary said that it was very dark near the fence and they couldn't discern how the creature hurdled the fence other than it went over it. As the creature was running, its arms were pumping, but not like a human would pump their arms; it was different. Mary said that the closest they got to the creature was probably 75 yards.

Mary McClelland's sketch of the creature. Confirmed by Inker.

Mary and Harvey went back and forth on the drawing until Mary told Harvey that he had it. She looked at the drawing and said that this was the creature that she and Inker saw on Highway 96.

During the interviews prior to starting the forensic sketches, we were constantly asked if people could look at the sketches completed for others. We never allowed anyone to look at any other sketches other than the one they were completing. The only exception to this rule was Inker McCovey. It was our general feeling that Mary had done a great job of doing her sketch. She wasn't overly eager to add finite details and readily admitted when she didn't see something or couldn't identify a specific point. If Inker was driving and she was a front seat passenger, she had the greater opportunity to concentrate on what was moving across the road and into an adjacent backyard. Inker had already completed two drawings and Harvey felt that he probably couldn't add much to what Mary had already described. We made a decision to show Mary's drawing to Inker.

Inker looked at the sketch and hesitated briefly while he examined the creature. With a slight smile, Inker looked at Harvey and me and stated, "That's what I saw." Inker said that Mary had done a great job describing the creature, and Harvey had sketched it perfectly.

Debbie Carpenter

Dental Office Assistant
Hoopa

Sighting #20: August 24, 2004

Debbie Carpenter grew up in the Hoopa Valley and spent her younger years hearing about Bigfoot and the havoc it caused in the Bluff Creek area. Debbie said that she remembers her dad telling her that Bigfoot was angry with the construction workers for coming into their area and building roads. She said that her dad and other tribal elders had told them to leave the Bigfoot alone, don't bother it and it will not bother you. Debbie said that she has always tried to take her dad's advice.

The Carpenters own a residence at the northern end of Hoopa that is downhill from the McCovey property. The Carpenters reside on a square of the valley that is immediately adjacent to the Trinity River, downhill and across from the sighting of another tribe member, Warrior Sanchez. Debbie said that her family has been fishing the same section of the Trinity for five generations. The streets that criss-cross the Carpenters' land all have huge

Debbie Carpenter standing at the location where she saw Bigfoot.

hedges of berries that act like barriers lining the streets. On almost any given day during the spring and summer you can find huge piles of bear scat that line the streets in the Carpenters' neighborhood, one of the few times I've ever seen something of this magnitude in a residential neighborhood.

In early April 2004 Debbie was working as a dental assistant and became engaged to Wendell White Sr. On April 22, 2004 at approximately 6:30 a.m. Debbie was driving back to her mother's residence on Carpenter Lane. Debbie said that she noticed that it was slightly foggy, with the fog sitting up off the roadway. She guessed that it was just after 6:30 a.m. and the sun was just starting to rise. She still had her headlights on. She was driving northbound on the main highway through Hoopa and found very little traffic on the roadway, less than usual. She turned westbound on Carpenter Lane at the southern entrance. She said that she was driving slightly downhill and towards the Trinity when she saw what she originally thought was a large bear standing on two feet. Debbie said that she was slowing to make a right turn to her residence when she looked straight down toward the river and saw the creature standing adjacent to a large blackberry bush. She notes that there was also a telephone pole next to the bush that she later used to judge the creature's size.

Debbie said that she was making eye contact with the creature as her car was moving. The creature turned slightly and then took two steps on the roadway toward the southern land plot. Debbie guessed that she was 100 feet from the creature as it was taking its steps. She said that she slowed down as she was making her turn, and got a good look at the size, shape and color of the creature. She stated that when the creature started to walk, she immediately knew she was looking at Bigfoot. She stated that it moved much like a human and she could tell that it was definitely comfortable moving and walking on its two feet.

Ms. Carpenter said that the creature she saw was black in color and had hair that hung down 6–8 inches on the forearms. She also stated that the creature was much thinner than the Bigfoot caught on the Patterson film. Debbie said that she would describe the creature as "skinny." She went back days later and measured that pole next to where the creature was standing and judged the creature's height at seven feet. She also stated that one major difference between the Bigfoot she saw and any man that may be in a costume is that her creature's arms hung much lower than any human arm. When the creature looked at her, it turned suddenly toward the pole. Debbie was unable to get a good look at its face, only a partial profile. The last she saw of the creature, it took a huge step over the adjacent berry bushes on the south side of the roadway and entered a field.

Immediately after the sighting Debbie drove directly to her mom's residence and told her what she saw. She stated that she also called her fiancé, Wendell "Winkle" White, 55 years old, and told him what had happened. Debbie signed an affidavit.

Update

Debbie was very willing to work with our forensic sketch artist and was looking forward to the experience. The week that Harvey Pratt was in town, Debbie's family suffered a death in Arizona and she had to leave to attend to family matters.

Location of Sighting

The location of the sighting is in the middle of Bigfoot Alley. This area is between the Mill Creek Road subdivision and Transmitter and Bald Hills, both areas with Bigfoot activity. Mill Creek flows into the Trinity River a few hundred yards south of this location and the activity on Bald Hills Road is just across the river and up the hill from the footprint location that her husband found. Debbie's sighting is in the middle of a large flat area of Hoopa that has huge berry bushes. The creature seen by Debbie later stepped over the bushes parallel to other hedges in an adjoining field to escape. This area of the Trinity is easily identified by several huge rock formations. These rocks can be identified from miles up Bald Hills and Mill Creek, which may be one reason that it appears to be a gathering place for Bigfoot. For a photo of the region refer to the Winkle White incident report.

I did clarify some of the wording used by Debbie in her description of the Bigfoot's escape. She did say that the Bigfoot "stepped" over the bush and crossed a field; she confirmed that this is what she meant. I later went back and measured the height of the bush adjacent to the pole at five feet, 10 inches. For any creature to have the ability to step over anything with a height and girth the size of the bush indicates the creature had to be huge. The pole and bush can be seen in the background behind Debbie in her photo.

Raven Ullibarri

Hoopa

Sighting #21: Fall or Winter 2004

Raven was referred to me by Inker McCovey. Inker stated that Raven had a Bigfoot in her backyard and had called the police over the incident. Inker and I tried to contact Raven the night of July 7, but she was not home.

Raven's home is the last one on Upper Mill Creek Road before it turns and goes up into the mountainside. Inker's residence is down the hillside, below Raven's. If anything were coming down the hillside, it

Raven Ullibarri at her residence.

would have to travel through Raven's backyard before entering any other yard. She is in an optimum location for her yard to act as a filtering mechanism for all creatures coming down from the Upper Mill Creek drainage.

I accidentally woke Raven up when I knocked at her door the next day. She was very polite and eager to answer my questions. She confirmed that she did have a Bigfoot in her backyard. She said that she thought it was in the fall or winter of 2004, she wasn't positive. She said that she was home and it was approximately 11:00 p.m. or 12 midnight. She was going out to a side-yard shed where she kept her garbage, and as she walked down her stairs she heard something near the shed. There were lights on near her garage and she could not quite see what was making the noise. She told me that all of her neighbors regularly have bear problems and it wouldn't be unique or scary to see a bear. She made it to the bottom of the stairs and saw a Bigfoot leaning over the door to the shed, as it appeared to be looking through the garbage. She said that it stood on two feet, was huge, had hair over its entire body, was over seven

Raven's residence with the garbage shed to the left of the garage where Bigfoot retrieved a bag.

feet tall, had very wide shoulders and didn't stop its rummaging when she made it to the driveway. Raven stated that she may have made a gasping sound because the creature did turn and look toward her. It had a dark coat and its face was a cross between a human and an ape, but more human. It was using its hands to look through the garbage in the shed as it was leaning over the side door. Raven said that the face of the creature did not have a snout like a bear, but it was more flat like a human.

Raven said that after she was in the driveway looking at the creature, she ran into the house, got her phone and immediately called the police. She knew that her boyfriend's brother was on duty (Sgt. Joe Masten) and he would respond. It was no more than two minutes after her initial call when Sgt. Masten arrived and seconds later Officer Joe O'Rourke pulled up. She ran down the stairs to meet the sergeant and saw the Bigfoot now slowly

Opposite page, top and bottom Raven's garage and garbage shed and the overgrown path that the Bigfoot took to the forest. She did say that weeds in the yard were much higher when the picture was taken than when the incident occurred.

Sgt. Joe Masten of the Hoopa Tribal Police, now deceased.
Thanks to the Masten family for the use of the photo.

walking with a bag of garbage toward her fence line and the forest. Raven said that the Bigfoot was carrying a white plastic bag in each hand as it made its way to the forest line. She stated that she turned as the first unit arrived and asked Sgt. Masten if he had seen the Bigfoot. He stated that he had not. She said that the sergeant must have arrived just seconds late.

Raven said that as she started to explain to Sgt. Masten what had transpired, Officer O'Rourke arrived. When she was giving a detailed account of what had occurred, all of them heard very loud and heavy steps. The steps were so loud that the earth felt like it was moving. She claimed that there were no more than 6–8 steps and then they stopped. The steps didn't fade in sound or intensity; they just stopped. At this point Raven said that Officer O'Rourke followed the route that the Bigfoot took and she could see him go lower into the property line as Sgt. Masten entered the forest at a higher point and went uphill. She said that Officer O'Rourke's canine was going crazy in the back of his police car, but was confined inside the car and wasn't leaving. Raven stated that it felt like the officers were gone for 30–45 minutes searching. She stated that both policemen had their guns out when they entered the forest.

The police returned, dirty and dusty, and told her that they hadn't found anything. She said that they couldn't believe they hadn't seen the creature. They both stated that they had only heard a few footsteps and they each couldn't believe that the creature was able to make an escape. They were baffled they hadn't found it. (Refer to Sighting #13 for Officer O'Rourke's statement.)

I asked Raven if she had any idea why Bigfoot would be going through her garbage. She explained that approximately four months earlier when she was having her menstrual cycle, she had several tampons in the garbage. In the middle of the night she heard something going through her garbage. She said it was a very dark night with no moon and no lights on outside. She looked towards her shed and saw an extremely large creature walk towards the fence as she started to walk towards the driveway. She said she thought this was also Bigfoot, but wasn't as upset because the creature left the area immediately. She also explained that the lighting wasn't working so she couldn't be positive on identification.

The Bigfoot on this later occasion was also going through the bag of garbage that contained tampons, because she was also having her period when this incident happened. Raven is convinced that Bigfoot can smell her tampons and is interested in the odor to the point of taking garbage that contains the items. She felt that the timing of the two incidents and the way the Bigfoot went for the bag of garbage with her sanitary items inside was too coincidental.

I asked Raven if there was any hair that was left behind after this latest incident. She stated that when Bigfoot was reaching over the barrier between it and the garbage, there was a clump of hair that was left behind on top of the left-side door that it bent over. She turned the hair over to another Bigfoot organization and according to Raven they had it tested for DNA. She states that the DNA came back as "no known animal or person." I asked Raven if she still had any of the hair leftover. She said that she thought she might, but wasn't sure as she gave most of it to another researcher.

Disappearance of Creature

In many sightings reports I have read and personally investigated, it is fairly common for people to claim that Bigfoot sometimes completely disappears. Some claim that they are following tracks and they suddenly stop, or they see the creature, start tracking it again and then it is gone. If we are looking at these reports from a scientific perspective, then common sense needs to be used in analyzing the statements. In the case of Raven's sighting, Officer Joe O'Rourke arrived shortly after the creature left the yard and he personally heard 6–8 footsteps, then nothing. An investigator has to ask where the creature can physically go. It cannot vanish. It cannot fly. It couldn't dig a hole and bury itself in that time frame. It could climb a tree. There is a grove of old growth in Raven's forest adjacent to her yard that is sufficient to hold an 800-pound mammal.

Officer O'Rourke made the statement that he felt that he was being watched while he was searching the forest for the creature. He did state he couldn't believe that it could have escaped the perimeter that he and the sergeant established. The proposed solution that I have developed is that the creature heard police cars arriving at the scene and heard the cackle of the dispatcher on their radios as they were standing in the yard talking with Raven. The creature took a few more steps and then started to climb. It heard multiple adults talking about the incident, more radios crackling and maybe saw guns drawn. There are huge trees behind Raven's house; some are 6–7 feet in diameter. Consider a statement made by Officer Greg O'Rourke (cousin of Joe O'Rourke) of the Hoopa Tribal Police, "Hoopa people believe that Bigfoot lives in the trees like in the movie *Predator*." If you take this statement and apply it to the Raven incident, then the act of Bigfoot staying in the area and avoiding detection has a rational solution. Many of the trees that are very large have huge canopies where a large animal could easily conceal itself without detection. The idea of the creature hiding in the trees makes Officer Joe's statement about feeling like he was being watched seem possible and believable.

Many months after interviewing Joe O'Rourke I was able to contact Sherrell Masten, mother of Sgt. Joe Masten. Joe died of a heart attack shortly after the Bigfoot incident at Raven's residence. I told Sherrell that I had seen an article in the Hoopa newspaper about Joe, and also saw a great photo of him next to his patrol car. I asked her if Joe had talked about Raven's incident. Sherrell stated that she could still remember Joe coming home and telling her that it was one of the eeriest nights of his career. She said that Raven was very frightened when the officers arrived. Joe got there in time to hear and feel the footsteps of the creature as it was leaving the area. She said that her son followed the creature into the forest and found it extremely odd that when he stopped walking, the creature stopped making noise. She said that Joe stated that the footsteps the creature made were very, very heavy. Sherrell said that she would never forget how her son described the Bigfoot incident that night.

Location of Sighting

Raven's residence is the last house on Upper Mill Creek Road and sits high above lower residences that sit next to Mill Creek. There are large hills that sit directly behind her house and a large field containing old growth trees that you can easily walk through to get to Lower Mill Creek Road. Her residence is a location that you would have to walk behind if you were coming off the hillside and wanted to get to Lower Mill Creek. It would not be hard to understand why a Bigfoot would wander into her yard and go through her garbage, based on the number of sightings below and around her residence. It can also be understood why a Bigfoot may have smelled the odor emanating from Raven's shed. The shed sits off to the side of the residence. The wind from the valley flows up into the hill and passes directly by the shed lofting the odor up into the hillside. If Bigfoot has a great sense of smell, then the odor could be sensed for miles. Raven's house is the start of the neighborhood, as you are traveling downhill. It should also be noted that Raven does have a powerful floodlight that sits above the garage near the shed. This light would have given her a fully illuminated view of the Bigfoot as it was rummaging in the garbage.

This sighting is highly unusual for a variety of reasons. If the Bigfoot had entered her yard months earlier, as she has explained, the creature has turned urban and is either feeding on human garbage or seeking used tampons. Raven's story about the tampons is not unique to this sighting. There have been several other sightings throughout North America where women have voluntarily stated that they were having their period when their sighting occurred, not a normal statement for most women to make under any condition. There have been a few sightings that indicated that Bigfoot was rummaging through garbage, but it hasn't been a normal occurrence in the Bigfoot world. To take a report where Bigfoot actually carried off two garbage bags is highly unusual, especially with the knowledge that there is not a lack of natural food in this region.

I did make an attempt to obtain the dispatch tape and 9-1-1 call on this incident, but found that each had been destroyed.

Forensic Sketch

Raven was very interested in completing a sketch with Harvey, and never wavered from her interest. I picked her up at her house and drove her to the hotel where she met the Pratts. We were all interested in this sketch because Raven probably had the longest look at Bigfoot of anyone I have ever interviewed. The amount of time it stood at the garbage and wasn't scared sufficiently to walk away is amazing. Harvey, Gina and I had agreed early in the meetings that we would concentrate on the facial structure with Raven and then possibly sketch a view of the backside of the creature as it walked into the woods. We explained our idea with Raven and she agreed that she would try.

Raven started the meeting with Harvey by explaining that she was approximately 20 feet from the creature at its closest point. She said that there was porch lighting and lighting from a motion-activated light on her garage. She said that the lights went on outside and she heard a noise near the garbage so she went out to investigate. She saw a creature that was approximately eight feet tall with a very heavy body frame. The creature was either dark brown or black, and had long hair coming from almost all parts of its body. It appeared that the creature did not have a neck, but it did have massive shoulders. Raven states that the creature did not emit an odor. The creature used its arm to shade its eyes from the light while simultaneously trying to hold two garbage bags in its hands. She guessed that the creature had stayed in her yard almost 20 minutes by the time she heard the noise, put on shoes, went down the stairs and the officers arrived in her yard. Raven also stated that she believed the creature's eyes almost glowed a yellowish or green tone. The creature left the area carrying one bag in its left hand (possibly another in its right hand, but she wasn't positive) and walking on two feet similar to a human, as it entered the forest in her backyard. She could hear branches and trees breaking, as

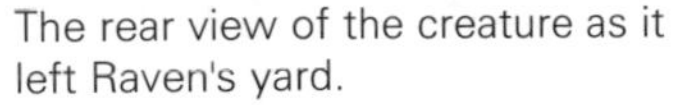

The rear view of the creature as it left Raven's yard.

The facial sketch done by Raven of the creature that visited her yard

it was trying to make its escape. Raven said that as officers arrived, the noise the creature was making would stop. When the officers started to walk, the creature started to make noise; when they stopped, it stopped again. Raven felt that the creature was a male because she did not see any breasts on it. She specifically noted the length of the creature's hair coming off the back of its head; it was 4–6 inches long.

Raven did collect hair left by the creature on the top of the shed it was leaning on. The hair was given to the Bigfoot Museum in Willow Creek. I later met with Al Hodgson who is the curator of the museum. I have purposely avoided being in the Bigfoot spotlight at media events, conferences, etc. There are a lot of interesting characters that are involved with Bigfoot, and I've found that staying away hasn't hindered my efforts. I must admit that Al Hodgson was a great person to meet. He has been involved in Bigfoot research since the late 1950s and early 60s, and was one of the first people that Roger Patterson called after filming the Bigfoot in Bluff Creek. Al has seen all sides of the Bigfoot phenomena, good and bad. In the early years he said that he wasn't a believer, but in the last several years, with the advent of technology and scientific advancement, he now knows that Bigfoot exists. Al agreed to meet with me at the museum in Willow Creek where he is a curator.

I got to the museum a little early and viewed the hair they have on file from Raven's sighting. It's quite unusual for hair, long (6–8 inches) and has a look like pubic hair. Al arrived a little after I did and we had a short discussion about the hair. He told me that he had a good friend who was a taxidermist in the area and Al allowed him to view the hair. The taxidermist had been involved with animals for over 50 years and had seen everything that had ever roamed the Six Rivers area. The taxidermist told Al that he had never seen hair from any animal that looked like this. He was positive that it wasn't from a bear, elk, moose, horse, cow, deer, nothing that he had ever worked on or seen, period.

I want to thank Al for his continued support and his great mind. He is someone that Bigfoot research needs, and hopefully he will be more in the limelight of Bigfoot lore in the years ahead.

When Harvey had finished the facial sketch of the creature Raven saw, he went onto the sketch of the back of it. Once Harvey had completed both sketches, Raven was almost speechless. She stated that the sketches look exactly like the creature, and she would change nothing about them.

Raymond Ferris

Hoopa

Sighting #22: October 2004

Raymond is close friends with Inker's family and was asked by the family to meet me at Inker's house for an interview. Inker said that the kid was reliable, honest and good-natured.

Raymond was with his mother, Diana Ferris, when this interview took place. Diana and Raymond live right next to Inker on the hillside portion of the property.

Raymond said that he thinks it was either a Friday or Saturday night in October 2004 when this incident occurred. He was out riding his motorcycle on the hillside road above their house, which they call Upper

Mill Creek Road. Raven Ulibarri's house would be the closest residence to the area where Raymond was riding. He said that he knew neighbors would start to get upset at the noise he was making, so he was trying to be respectful and was gliding down Mill Creek Road toward his house with the engine off. He said that his cycle did not have a light, so he was carrying a flashlight in his hand. He reached the area above Raven's house where there was a slight turn and he started to slow. He said that he saw a large object on the right side (his side) of the road. He initially thought it might be a bear because of how large it was and its dark color; bears are a very common sight in Raymond's neighborhood. He said that he tried to stop the cycle as fast as he could in order not to scare the animal or have it run into his path. He said at this point he was just above his house, maybe 300 yards through a large field filled with trees.

Raymond Ferris

Raymond states that he couldn't get the cycle to stop fast enough and got uncomfortably close, approximately 10 feet from the creature. He said that he pointed his flashlight at it and saw it was not a bear. He said it had a human-type face, with dark hair over its entire body. He said that creature was kneeling on the shoulder of the road with its elbows leaning on its legs. The entire time he was riding towards the creature, it never moved; it was completely motionless (sounds familiar). He said that the creature's reddish-yellow eyes were penetrating and frightening. He states that a few seconds elapsed and the creature stood up, turned to the hillside and casually walked up the embankment and into the woods. It was spreading trees with its arms as it went right through the middle of a grove. Raymond states that the creature was huge, guessing it was between seven and eight feet in height and hundreds of pounds in weight. The one additional feature about this sighting that was slightly unique was that Raymond claimed the creature stank very badly. He said that it smelled like a very dirty, wet dog. He said that he remained stopped on his cycle watching as the creature made its way up into the hillside, and couldn't believe what he had just witnessed.

Raymond went straight into his house, locked the doors and went to bed. He said that it was really late and he didn't want to get in trouble. He did tell his mom the next day, and she stated that she had smelled that stinky smell as she was going to bed that night and distinctly remembered how horrible it smelled.

I met Raymond several months later at the local teen center. He was boxing a fellow tribe member in a training regime. He took time out to talk with me about his sighting and his activity in recent months. He was very polite and absolutely certain of what he saw in October 2004 on Upper Mill Creek Road: Bigfoot! He said that there was no doubt in his mind that he witnessed a Bigfoot at 10 feet. He said that it walked much like a man, but was much too large for any human. He explained that the stink that came from the creature was horrid. He felt that he was very lucky to have seen the creature.

Location of Sighting

Raymond's sighting occurred several hundred yards up the hill from Raven Ullibarri's residence on Upper Mill Creek Road, also in Bigfoot Alley. This area has had several sightings in the 2003–2005 time frame. In the area of this sighting, the road is a dirt–gravel mix and goes into the wilderness area behind the reservation. This is a well-traveled road during daylight hours because of the timbering and roadwork done by tribal forestry. I have personally traveled this route several times and am amazed at how the countryside changes as you meander your way throughout the forest. It changes from heavy timber to chaparral to alpine. If you take this road long enough you can actually make your way to Orleans through a backside mountain route. It would take a United States Geological Survey topographic map to find your route, but I have done it. Reminder: you are not allowed on dirt and gravel tribal roads without a permit or you will be arrested.

Raymond signed an affidavit.

Comments

A review of Raymond's description of the incident will give some insight as to how he happened onto the Bigfoot. First, Raymond was obviously concerned about waking neighbors and thus had the motorcycle turned off. If he was coasting down the hillside (as he claimed) the animal may not have heard him. It could not have seen him coming because he was around a turn and up the hill from the creature. Out of the hundreds of sightings I've reviewed and read in the Northern California area, it is not uncommon for Bigfoot to stand completely still when someone is walking/driving up on it. If there is something to stand behind, the creature will do it. If there is no cover nearby, the creature will sometimes crouch (as it did in this incident) and hope that the witness continues past without it being noticed. The escape route that the creature took should be noted. It walked calmly uphill from Raymond, and not downhill towards Mill Creek. It might have taken this route because it would have had to walk around Raymond to go downhill, or it might have taken the same route up as it did coming down. There are berry bushes very near this location, as well as nuts, mushrooms and acorns. The time of the sighting is also consistent with the Inker McCovey and Mary McClelland sighting on the highway directly below this location, 11:30 p.m. to midnight.

Something that is becoming obvious to this researcher is that the Bigfoot's sense of sound may not be one of its strong senses. Many of the sightings of Bigfoot have occurred while vehicles drive up on it, usually downhill. The creature appears surprised and escapes by going down the hillside. There are also many sightings where witnesses see Bigfoot on a trail, and if the wind (sense of smell) is blowing towards the witness, they can sometimes walk right up on the creature without being heard. It has been in the last several years that I've become especially cognizant of this deficiency in the creature's senses.

Forensic Sketch

Raymond was one of a few people who got very close to a Bigfoot. He was someone that we had wanted to attempt a forensic sketch. Unfortunately,

Raymond was out of town the week we were conducting the sketch interviews, and wasn't able to participate.

Michael Mularky

Store Manager
Hoopa

Sighting #23: November 2004

Michael Mularky grew up in Northern California and heard the stories of Bigfoot, saw footage taken of the creature and heard local Hoopa stories of it living in the area. Michael never believed the stories and always felt they were someone's imagination gone wild.

Michael is a 30+-year-old store manager at Ray's Supermarket in downtown Hoopa. Michael talks to the local people daily, listens to their stories and talks about their folklore. He has heard stories of creatures in the river and Bigfoot in the mountains. Michael is a very hard-working man who leaves his house in the early morning hours every day to ensure the stores shelves are filled when the customers arrive at 7:00 a.m.

Michael lives in the small town of Willow Creek, approximately 14 miles south of Hoopa. Willow Creek claims it is the Bigfoot capital of the world and boasts one of the few Bigfoot museums. The town is small, one main street, and has a few shops, restaurants and gas stations. The Trinity River runs along its perimeter and offers a beautiful backdrop to a quaint town. The city is surrounded on all sides by huge mountains that are covered with lush vegetation and huge trees.

On a cold morning in November 2004, Michael started his day with the normal routine of waking at 3:30 a.m., showering, dressing and getting into his cold car for the half-hour drive to Ray's Supermarket in Hoopa. The drive is a leisurely trip along the banks of the Trinity, passing the sheriff's substation, the forest service office and the California Highway Patrol station. The road meanders its way through a cliff area with a steep drop into the river. The road also passes a campground, Tish Tang, which is located on the banks of the Trinity. Years ago there was a reported Bigfoot sighting at Ishi Pishi, but the reports on the west side of the river have been rare.

Michael in front of Ray's Supermarket.

Michael was approximately 15 minutes into the drive when he entered Hoopa and an area of the highway that parallels Shoemaker Road. It's a region known to him for having a small computer shop on the west side of the road and an espresso location on the river side. He said he had his headlights on and was traveling approximately 40–50 mph when he saw a huge creature standing on the roadway near the computer shop. He immediately slowed when he saw the hairy beast standing on two feet. Michael said the creature was covered in reddish-orange hair or fur, except under its arms, and was standing on two feet. He stated that the creature did not have a snout like a bear, but had a flat face like a human or ape. He said he saw the creature take two giant strides, 6–8 feet each time, as it walked across the roadway and attempted to partially hide behind a large tree on the eastern side of the road. Michael said that he continued to slow his vehicle to get a good view of the creature. He said that it could not get completely behind the tree, and appeared to be looking at him as he continued driving. Michael said that as the creature was looking at him, he could see that its eyes seemed to be almost glowing yellow. He continued his journey to Ray's and told a few friends about what had happened later in the morning.

Michael said that what he saw that morning in November 2004 was definitely a Bigfoot. He said that it matched the descriptions of some of the photos he has seen of the creature. He said that the creature always walked on two feet, was huge, covered in hair, took large strides, was 7–8 feet tall and obviously was very shy about being seen. Michael said that he was the ultimate skeptic about Bigfoot being a living mammal, but no more. He stated, "I know what I saw and it was not a bear, it was Bigfoot. Those eyes were unreal, I'll never forget its eyes or its size." Michael signed an affidavit.

Location of Sighting

The location of Michael's sighting is another spot I routinely pass by on my way into Hoopa. It is located on the main road into and out of Hoopa and runs along the west side of the Trinity River. One of the main factors that is significant about this location is the abundance of berry bushes on both sides of the road and along property lines in the area. It would appear when Michael saw the creature it was probably feeding on one of the berry bushes. This location is also in close proximity to Tish Tang Creek and the area where sightings have been made on the eastern side of the Trinity River and along Shoemaker Road.

The area of Michael's sighting is also in the same general proximity as Hank Masten's, Corky Van Pelt's incident, and the Michelle McCardie and Leanne Estrada sighting. These are all very close to each other.

A unique item about this sighting is that Michael is not native to the area, is not a Hoopa tribal member, and was a skeptic about the existence of Bigfoot. He is well-educated, very hard working and an absolute straight talker. Michael never wavered in his statement and never hesitated about his answers. It would have been very interesting to see what Bigfoot would have done if Michael had immediately pulled to the side of the road adjacent to Bigfoot. We will never know.

Michael has always been very steadfast on his sighting. I think he felt bad because he questioned what others had said about Bigfoot until he was a witness. He said when it was his time to assist in sketching the creature he would gladly help.

Michael had a hectic day at work before his sketch appointment, and was a bit nervous coming into the hotel room. Michael started the session by briefly explaining to Harvey that he was a non-believer in the creature until he saw it. He stated that it was 3:00 or 4:00 a.m. and he was driving into work at the supermarket. He came upon an area near Wanda's Computers and saw the creature standing near the road. He said it was standing on two feet, had hair over its entire body and it was huge. He guessed that it was near eight feet tall and built very solidly. He stated that the arms were longer than a human's arms and they swung when it walked. He said that the lights from his vehicle illuminated the creature and he could tell that it didn't like the lights, as it turned slightly away. The creature had 10-inch-long brown hair streaming from the back of its head, and shorter hair on its facial area, which Michael could see flowing while the creature was moving off the road and onto the shoulder. Michael stated that he feels he observed it for approximately one minute at a distance of close to 100 feet. He saw the creature move to an area on the shoulder of the road behind a tree and attempt to hide. He also felt that the creature appeared dirty even though it looked healthy and big. Michael said that he would think that he was looking at a male.

Michael Mularkey's sketch of the creature that he saw cross the road in south Hoopa.

Once Harvey was done with the drawing he showed the sketch to Michael. Michael stated that it was the exact creature that he had seen cross the road that morning. He said that he could never drive through that location again without thinking about that giant creature and its eyes.

Michael never saw any of the other sketches completed by our witnesses. In our review of the sketches by Raven Ullibarri, Kim Peters, Jackie Martins and others you will notice a very similar physical appearance. The creatures all appear to have similar physical builds, which lends credibility to the stories and the ability of Harvey to draw the descriptions from the witnesses.

Gordon McCovey

Hoopa

Sighting #25: May 2005

The winters in Hoopa can be very wet, rainy and sometimes depressing. In one of my many trips during 2007 I had hit a stretch of weather where is seemed to rain for days, no sun. During those rainy days I would catch up on interviews with local officials, spend time in the mountains in my vehicle and finally catch up on my writing. It was one of those rainy writing days that I had a knock on my hotel room door. Just as I was walking to the door the sun started to shine through the rear sliding glass door and I thought to myself, this might be a good sign.

I opened the door and found a very slender, tall and dark complexioned gentlemen standing in the hallway asking if I was Dave. I stated that I was and he said that he had a Bigfoot story that he would like to talk to me about. I asked him to meet me downstairs and I would gather my materials.

The individual introduced himself as Gordon McCovey. He stated that he was 53-year-old lifelong Hoopa tribal member and he had heard from friends that I could be trusted if he told me about a Bigfoot sighting. Gordon stated that he had attended Hoopa High School and had spent the majority of his life working and living in the valley. He said that the farthest he had ever traveled from the reservation was Montana. He explained that he had held jobs for the tribe in timbering and logging, and was presently disabled. He said that he tried to get out and walk in the woods as much as he could and has attempted to enjoy the scenery and beauty that surrounds Hoopa. The walks helped clear his mind and offered him a sense of freedom.

Gordon explained how he had heard stories about Bigfoot his entire life, but had never seriously thought he would see it. It was mid-May 2005 when Gordon and a female friend, Moonshade Dowd, decided to go mushrooming together. Gordon said that they were going for morel mushrooms in a burnout area just east of the Dillon Creek Campground east of Somes Bar.

Gordon said that he always liked to get an early start when he goes mushroom hunting, and that day in May 2005 was no exception. They left his friend's house at close to 5:00 a.m. and drove straight to the location. They parked their car in a small turnout on the main highway; it was approximately 5:30 a.m. Gordon said they sat in the car for several minutes just talking and enjoying the scenery as the sun started to rise. He stated that they were both looking across the Klamath River when they saw a huge creature appear from behind boulders on the opposite bank. The creature was walking on two feet at a brisk pace and swinging its arms as it moved. Gordon said that its legs were slightly bent, it had hair over its entire body, and it moved much more like a human than animal. He described the creature as having hair or fur over its entire body and its color as dark brown.

Gordon said that the creature was walking on a combination of sand and rock, and for a period of time didn't notice that they were watching it. Gordon said for some reason the creature looked in their direction, obviously saw their vehicle and immediately picked up its pace. Gordon states that at the closest point they were approximately 200 yards from the creature and at the farthest were close to 300 yards. The creature continued to quicken its pace until it reached the forest edge and then disappeared.

Gordon said that he has seen many of the film clips of Bigfoot and the Bluff Creek footage. He stated that the creature moved in almost an identical manner as the creature in that film. The creature's appearance was very close to the one at Bluff Creek, but he couldn't tell if it was male or female.

McCovey said that as he and Moonshade watched the creature disappear into the forest, he explained how lucky and shocked he was that they were able to see it. He states that Moonshade initially felt that he had staged the entire incident. He explained that there was no way anyone

Gordon McCovey.

could get onto that opposite bank from anywhere near where they were located, and no sane person would get up at 6:00 a.m. to pull a stunt like that. Gordon said that Moonshade understood and was still in shock that they had seen the creature.

Gordon said that he is 100 percent positive that he and Moonshade saw Bigfoot that early morning in May 2005. Gordon signed an affidavit.

Location of Sighting

This location fits nicely into the portfolio for Bigfoot sightings in and around Orleans. The Dillon Creek Basin has long been a place where researchers have studied the Bigfoot phenomena. The area of Dillon Creek upstream from the Klamath is very rugged. In the past, forestry crews have become lost in that region, an unusual thing to happen to a seasoned crew, which speaks to the isolation and rugged nature of the region. The headwaters of Dillon Creek are located on an opposite ridge from the headwaters of Bluff Creek from the Go Road. The Dillon Creek area is much more rugged and less traveled than the Bluff Creek Region, and is an area with far less foot traffic and fewer human visits than Bluff Creek today. This area rates very high for future study. It is off the radar of most Bigfoot enthusiasts, and yet should be a location that receives more interest. If you refer to the comments made by Tony Hacking when I asked him what areas would be high on his list for the possible study of Bigfoot, one location he mentioned was the Dillon Creek Basin.

Paul James

Richard Nixon

Fishermen
Hoopa

Sighting #26: April or May 2005

Paul James is a close friend of Inker McCovey. Inker first took me to Paul's house the first day I met him. Paul was very polite, but stated that he was exhausted from working a long day and asked Inker if I could return the next morning, which I did.

Paul is a 59-year-old Hoopa tribal member who lives approximately one-quarter mile up the road from Inker, and about a block downhill from Raven. Paul lives on Upper Mill Creek Road and Inker lives on the lower portion. Paul has a hill sloped downhill behind his house that leads to the local highway and eventually to Debbie Carpenter's house and the Trinity River — all in Bigfoot Alley. Paul's residence is located approximately 50 feet from the Bigfoot sighting of Kim Peters.

Paul has had two specific sightings of Bigfoot, one with two witnesses and one alone. As the interview started, Paul's 18-year-old son was in the room along with Paul's girlfriend of 25 years, Leslie Abbott.

Paul's first sighting was in April or May 2005 approximately 12 miles

Paul James inside his residence

downriver on the Klamath from Weitchpec. Paul said that it was close to dark, maybe 8:30 p.m. when he, his son-in-law Richard Nixon, and brother were returning from a day of setting nets in the river. (Tribal members have been given the right by the federal government to net salmon and steelhead from the Klamath and Trinity rivers.) He said that they were driving up Highway 169 adjacent to the river and were on the river side of the road. He said that he was driving, with his brother sitting in the front seat of the truck next to him and Richard sitting in the middle rear area of the cab. He said they were traveling approximately 35 mph and had just made a turn. They could see that approximately 100 yards down the road something very large was coming down the embankment very fast. His first thought was it must be a bear because he could see it was large, shiny black in color, and coming from the woods. Paul said his thoughts were only fleeting because he quickly realized that the creature was on two feet, not four. He said that the creature also had very long arms that were swinging almost in unison as it was making its way downhill. He said that the creature's legs were bent as it was running down a very steep embankment, and then it jumped almost the last 15 feet onto the roadway. Paul said that his headlights were directly on the creature at a distance of 100 feet. He said that no man could move as swiftly down such a steep slope and jump in front of his truck as the creature did without being seriously injured or killed.

Paul states that the seven-foot-tall creature stood on the road just briefly, very briefly, and then took two huge strides across the roadway and jumped completely off the riverside embankment down towards the Klamath River. Paul was shocked that the creature jumped off the roadway because the hillside leading to the river is so steep. Paul said that everyone in the truck was speechless and really didn't say anything except stare at each other immediately after the incident. They drove straight home after the sighting.

Leslie Abbott was home when Paul and his relatives arrived back at the house. She states that in the 25 years she has known Paul, she has never seen him so excited and "alive." She said that it was obvious that Paul, his son-in-law and brother had just seen Bigfoot.

Paul said that he has seen hundreds of bears and other wildlife in and around the reservation, and he has never seen anything like the creature that jumped in front of his truck. He claimed to have seen videos and photos of Bigfoot, and the creature he saw in front of his truck matches the pictures and video of Bigfoot seen on television.

Richard Nixon's Account

In July 2006 I met Paul James and heard about his sighting west of Weitchpec. I found Paul's sighting valuable because there were two other men with him in the car when it occurred. I also found Paul very credible and was quite interested in meeting the other men with him. It took me five months to eventually catch up with Paul's son-in-law, Richard Nixon.

I had called Paul earlier in the week and he invited me to his residence so he could review his statement and sign the affidavit. Paul also advised me that Richard would be with him and I could interview and get his rendition of the sighting. Paul said that they were both working together doing brush clearing above the reservation and would be home early in the afternoon.

I met Paul and Richard as they pulled into Paul's residence on Upper Mill Creek Road. Because of the number of sightings on Mill Creek I am always observant when in that neighborhood, and since it was getting dark earlier than the summer, I was especially on the alert, but nothing was to be seen.

Richard exited Paul's truck and immediately walked up to me with Paul standing next to him. Richard introduced himself and had a very confident, positive approach that immediately struck me as a credible person. Paul gave me a big smile and handshake and said that he had missed seeing me over the summer. Paul and his family had been working in Eureka during June, July and August while I was at the reservation. I had left him several messages, but we always seemed to miss each other. Paul invited me into his house and allowed Richard and me to sit and talk.

Richard stated that he was a Hoopa tribal member and was the son-in-law of Paul. He said that they often fished together in the past and spent many nights driving along the Trinity and Klamath Rivers on their way home. Richard explained that he was 42 years old and had heard many stories over the years about Bigfoot and locals claiming to see it. He said that he was always polite when listening to others stories, but always thought in the back of his mind that they might be seeing a bear or they might be fabricating the story.

Richard said that it was in April or May 2005 when he, Paul, and Paul's brother were traveling eastbound towards Weitchpec from down river on the Klamath River. He said that they had spent their entire afternoon setting nets and attempting to catch salmon. He stated that Paul was driving and Paul's brother was in the right front seat and he was in the middle of the backseat with an unobstructed view through the front windshield. I asked where Paul's brother was now. Richard said that he was in a convalescent home in Eureka, seriously ill.

Richard claimed they were roughly 12 miles from Weitchpec at approximately 8:00–8:30 p.m. They had come around a small turn and had their headlights on when they saw a huge, dark-colored creature coming down the embankment on their left. They were traveling approximately 35 mph, according to Richard. He said that he could tell by Paul's slowing of the truck and head movement that he had also witnessed the creature. Initially, Richard states he thought he was looking at a huge bear, but those ideas quickly went away.

Richard said that when he first observed the creature coming down the embankment he thought that it was moving on four legs. After watching the creature longer, Richard said it was obvious that the creature was almost running downhill on two legs and then leaping the final 15 feet to the roadway. He watched the creature land on the roadway in front of their vehicle and stand on the road on two legs. What immediately struck Richard as highly unusual was the size of the creature's shoulders. He stated the shoulders were huge and the creature's upper body tapered to the waistline. The creature was covered with dark-colored hair/fur and stood erect to a height of almost seven feet and standing like a human. Richard further explained that the creature had a face more like a human or gorilla and did not have a snout like a bear. The creature's body reminded Richard more of a well built human athlete than a bear. The creature stood on the roadway for a few seconds and then leaped over the shoulder of the road and down towards the Klamath River. Richard said that he got an excellent view of the creature and is 100 percent positive he saw Bigfoot.

Richard Nixon at Paul James's residence.

Richard explained that he had heard hundreds of stories of Bigfoot over his years growing up in Hoopa. He said he has seen specials on television about Bigfoot and has seen clips of the Patterson–Gimlin movie, and stated that in general this is how the Bigfoot appeared to him. He didn't see any breasts or male genitalia. During the entire ride back to Paul's residence Richard stated that they were all in a state of shock, but they all agreed that they had seen Bigfoot. I did brief Richard and Paul about the number of sightings in their immediate community. This information made them much more comfortable about going public with their sighting.

Some Native people don't want to admit they saw Bigfoot, some are eager to talk about it and others brag about it. Richard hasn't told many people because he is afraid of being ridiculed. He does say that he feels he has been blessed to witness the mammoth creature. Richard Nixon and Paul James each signed individual affidavits explaining their sighting and they had each agreed to the basic facts of the sighting.

Leslie Abbott (Paul's Girlfriend)

I am including a short segment in Paul's sighting about his girlfriend and Kamoss. Kamoss is one of those issues that keeps "raising its ugly head" (in a matter of speaking) and there are many people who have approached me about the creature. The story doesn't seem to die and the creature has apparently lived for hundreds of years. This story goes to the oddity of the area and the possible fact that this region of California probably holds many more secrets than outsiders realize. I know this is off the direct Bigfoot agenda, but then there are some who believe that Kamaoss is really just Bigfoot swimming in the river.

Leslie stated that she has lived with Paul for 24–25 years. She said that a lot of strange things occur in their valley. She reminded me that the Hoopa valley is one of only two valleys in the Northern California mountain range, the other being the Scott Valley near Fort Jones and Etna, which is east of Hoopa.

Leslie said that there is an old Indian legend about a serpent that lives in the Trinity/Klamath River systems called Kamoss. She said that Kamoss can be as long as 18 feet, has a horse-type head and travels the length of both rivers all the way to the ocean. In the winter when the waters are high the serpent is in the river, and when river is low in the summer it retreats to the ocean or tidal areas.

Leslie claimed that in August of 1998, she, Paul and Paul's brother were out working their nets in the Klamath River approximately 18 miles down from Weitchpec. She said that is was light outside and near 8:30 p.m. Paul and his brother were approximately 100 yards downstream from her when Paul said that he heard her screaming. Paul said that both he and his brother started to move in her direction.

Leslie said that she saw Kamoss stick its head out of the water and slowly move downstream in Paul's direction. She stated that she was standing near Moore's Rock. The river is very deep in this area and if a creature were to live in the river, this would be a perfect place. She said that she watched Kamoss for approximately 40 seconds before it completely concealed itself in the water. She could see its body at the water line behind it, maybe for 10 feet. It acted like a snake that was swimming down the river, except its head was the size of a horse. She stated that the head protruded out of the water approximately one foot.

Paul validated that Leslie was hysterical, but that he and his brother didn't see anything. He also stated that Leslie told them to get out of the water.

Paul has been netting fish on the Klamath and Trinity River for almost 45 years. He said that in those times he has had nets stripped to a five-foot diameter. He explained that there are large sturgeon in the rivers, but the sturgeon would rarely tear five-foot gaps in the nets because there aren't that many huge sturgeon. He said that even a huge sturgeon wouldn't tear apart his nets to that extent. Both Paul and Leslie stated that over the years many people said they have seen Kamoss, but they don't personally know anyone who has ever seen it. Kamoss is something that is not generally spoken about at the local café.

Paul stated that ever since the night Leslie saw Kamoss, she won't go back to Moore's Rock. He has prodded and begged her to go back, but she won't go near it! It's obvious that something in the water disturbed Leslie. The real question: was it Kamoss or Bigfoot?

Kamoss is a creature that I have heard about from many credible Hoopa people. The alleged size of the creature, its ability to move swiftly through the river, and its possible proximity to nets arouse suspicion. I mention it here so readers can make a mental note, as it will be discussed in a later chapter.

Location of Sighting

The location of the James–Nixon sighting fits geographically with my knowledge of the area. On the south side of the road are the Bald Hills Mountains, a classic location in Bigfoot hunting and a location where three sightings are documented in this book. There have been many sightings in that area and it fits because of its desolation, lack of residence and no paved roads. This would be on the coastal side of the river and is approximately 20 air miles from the ocean. On the mountain side of this sighting is the absolute classic Bigfoot location, Bluff Creek. Bluff Creek is located 10 miles back into the Onion Mountains and is famous in Bigfoot lore for the Patterson–Gimlin footage. If you were going to pick the perfect location to sit on a road and watch for Bigfoot crossings, this might be it. The Klamath River has a plentiful food supply of salmon and steelhead, and is an obvious water supply. The river could also be viewed as a travel route for Bigfoot should they want to quickly, efficiently and in a stealthy manner move from the mountains to the ocean. J. Robert Alley writes in his book *Raincoast Sasquatch* (Hancock House Publishers) that Bigfoot has been seen many times along the Alaska coast swimming in the ocean between islands. "The ability to swim powerfully underwater is also noted in old Haida and Tlingit stories." It is reportedly an excellent submerged swimmer with its legs mimicking a frog's movements in the water. If Alley's reports from Native American tribes in Alaska and from fisherman are correct, then there is a significant likelihood that Bigfoot in California could also swim and would use the Klamath and Trinity Rivers as a means to move through the area.

One interesting aspect to this sighting was that Paul had complained to me in the past about his nets being destroyed by something huge. He stated that the nets are very strong and there has been something damaging them. Paul felt that he wasn't the only one suffering the loss and that other Hoopa people have also been victimized. He stated that the general tribal consensus was that giant sturgeon were getting caught inside and destroying the nets in an effort to escape. Maybe something else is destroying the fishermen's nets. I discussed with Paul the possibility of Bigfoot taking fish from the nets. He said that he really never thought about it, but it made perfect sense. He said that some of the areas where the nets were vandalized and fish taken were not near the banks of the river and it would take a small boat to reach them. If Bigfoot could swim (we know it can) then it would follow that it could go into nets in the river, take the fish and accidentally destroy nets in the process.

Pliny McCovey

Hoopa

Sighting #27: July 2005

Pliny McCovey is an uncle of Inker McCovey and lives on Upper Mill Creek Road in the center of Bigfoot Alley. Raven Ullibarri lives just

Pliny McCovey standing in his backyard in front of the brush where he made his Bigfoot sightings.

below. Pliny is another individual who has had Bigfoot sightings. He is older than many of the witnesses, but brings with him a sense of sincerity that lends itself to credibility.

Pliny told me that his most recent incident occurred in his backyard, and he invited me out to view the area. We went into the yard and he began to explain his sighting. Pliny's house sits on one acre on Upper Mill Creek Road with another residence just up the hillside, maybe 50 yards away. There is nothing behind Pliny's house except an empty hillside with a lot of brush that adds a very thick cover. There are several large trees that offer shade and obstruct the view over the embankment. The ridge where Pliny lives is the same ridge that gradually descends into the area where Debbie Carpenter resides, all inside Bigfoot Alley.

Pliny explains that one warm summer day, in the early evening, his windows were open and he heard noises in the backyard. He said that he went out and slowly made his way to the thick brush that surrounds the perimeter of his yard. He pointed out some of the large trees and grass that proliferate in the area. He said that the sounds weren't like anything he had heard before. He said they were a high–low whine or wail. The sound fluctuated to the point where it almost sounded like two animals. He moved very slowly and quietly towards the brush until he could almost hear the creatures breathing. The underbrush was fairly thick, yet he could see one large, hairy creature standing on two legs approximately 20 feet downhill from his location. Pliny showed me where the creature was standing and, based on the hillside angle, the creature appeared to be approximately 7–8 feet tall. He states that the creature he saw was covered with hair/fur, stood upright the entire time he watched it, and did not have a snout like a bear; the face was more human or ape-like. He said that he could hear something similar to a language that he couldn't understand being communicated between the creature he could see and another that was concealed. Based on the volume and tone of the communication, Pliny feels that the creature he couldn't see was smaller, maybe a baby Bigfoot, and the one he could see was the parent.

Pliny said that he kept extremely quiet and continued to observe what he thought was continued communication between the creatures. He stated that he couldn't see the second creature so he took another position further up his property. At this point the larger creature heard him and immediately made a sound, and Pliny heard both creatures start to walk down the hillside. He stated that it wasn't a rapid movement, but he could tell that they were moving out of the area. Within minutes the creatures were out of sight and the sounds stopped.

The exact location where Pliny saw these creatures is important to understand. His backyard is part of a declining ridgeline of the mountain above Mill Creek, and is the ridge that eventually falls and disappears at Carpenter Lane. If you followed the path behind Pliny's house and behind the bushes, the path would lead directly to the area of the Kim Peters sighting and into the area of the Debbie Carpenter sighting. If you went uphill you would end up across the street from the Raven Ullibarri sighting. By connecting the dots you will see that all of these sightings have a common thread, and yet by the descriptions (Debbie Carpenter saw a skinny creature, Pliny saw a large one, and Raven saw a huge creature) there have to be several different creatures in the area.

Pliny said that the next day after the sighting he went further into the backyard trying to determine what the creatures might have been doing. He stated that there is a small berry bush where they were located and it appears they were eating berries. He told me that he could not find any footprints and didn't see any scat. He did note that in the three months just prior to this incident, his family had lost three cats. He said that the cats had completely disappeared and he can't say Bigfoot took and ate the cats, but he finds the timing of their disappearance and the sighting of Bigfoot suspicious.

Pliny said that he knows that neighbors in the Mill Creek area have all had visitations similar to his. He explained that Bigfoot is understood by his people as the keeper of the mountains, and a friend. Pliny signed an affidavit.

Location of Sighting

Pliny lives in an area of Upper Mill Creek Road that is surrounded by Bigfoot sightings. Below him there have been sightings on an adjacent road, and as you move closer to the Trinity River there are additional sightings. Up the hill from Pliny is the location where a Bigfoot was observed going through a neighbor's garbage (Raven Ulibarri) and across the roadway towards Lower Mill Creek are a series of sightings (Masten and McCovey). The ridgeline where Pliny resides starts in the Upper Mill Creek basin and continues until it reaches the flats of Carpenter Lane.

The time of this sighting is a little unique. Most of the sightings in this neighborhood have occurred late at night or very early morning. This sighting occurred in the morning during daylight hours. It is also unique because two Bigfoot were supposedly simultaneously in the area. The timid nature of Bigfoot is exemplified in this sighting as it fled upon being revealed. I had a good opportunity to investigate the hillside where Pliny observed the Bigfoot. The area is steep and has significant cover. Because of the angle of the morning sun and the thick cover, the area would have been quite shady and cool, another ideal location for Bigfoot. The area also gives the Bigfoot an easy escape route through the bushes and up the hill into a very remote mountainous area.

Forensic Sketch

Pliny was one of the witnesses selected to assist in completing a sketch of the creature they witnessed. I went to his residence and spoke to him about the project and he agreed to do the sketch. On the day that Pliny was scheduled, he stated that he didn't feel comfortable, and politely declined and apologized.

Kim Peters

Pre-School Teacher
Hoopa

Sighting #28: August 2005

Kim was at her residence when Inker McCovey called her and asked to go to Pliny's residence and talk to me. Inker told me that Kim was embarrassed about her Bigfoot sighting and afraid nobody would believe her. She was slightly reluctant to file a report because of public ridicule, but after hearing about the number of sightings that have actually taken place in her community, she agreed to go public.

Once Kim arrived, she advised me that she was a pre-school teacher for tribal members. She said that she lived just down the street from Pliny and had lived in the neighborhood for a few years.

Kim said that approximately 11 months earlier, sometime in August, she was driving home from Weitchpec at 11:30 p.m. She took the shortcut up to her residence from Lower Mill Creek Road that meets the highway. She explained that this road is very dark and has no streetlights or houses until you reach the turn at Paul James's residence. She stated that she had her high beams on and was traveling close to 25 mph. As she slowly reached the top of the hill she saw something extremely large on the right shoulder of the road. The creature was standing on two feet, was over seven feet tall, had huge shoulders, and had hair over its entire body. She said that she knew it was not a bear because it did not have a snout like a bear, it was standing on two feet, had long arms and it had the body more of a large human than a bear. Kim said that it was a Bigfoot. Kim said that she has seen specials on television about Bigfoot and has seen photos in the press of the creature. The animal she saw on the roadway in August 2005 matched the description of the Bigfoot she has seen on television and in photos. She stated that the creature slowly moved into the trees and disappeared. She drove the short distance to her residence, went inside and closed and locked the doors. Kim said that she has been careful to whom she tells her story because she doesn't think many people will believe her; it was hard for her to believe what she had seen. She stated that the total time for the sighting was approximately 15–20 seconds.

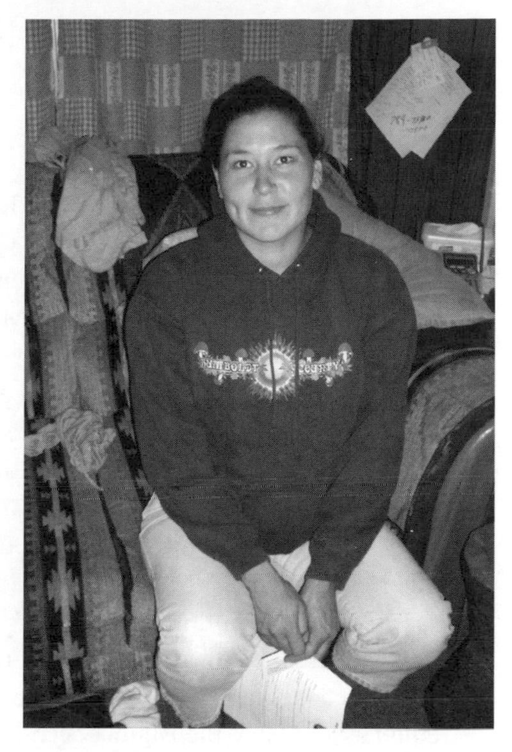

Kim Peters the night of the interview.

Kim was very careful with her words, and wanted to be extremely accurate with her descriptions. She was very articulate, easily understood and extremely polite. Kim signed an affidavit.

It should be emphasized here that Kim was a Bigfoot witness we found, she didn't find us. She was deeply afraid that people would make fun of her, tell her she was crazy and say that she just saw a bear. It took both Inker and me to convince Kim to be comfortable and confident enough to come forward and tell her story in a public forum. The concerns Kim expressed exist with many Bigfoot witnesses and goes back to the same concerns expressed by victims of crimes. It's hard enough to witness the event, but it's worse to be ridiculed for just telling people what you saw. If it wasn't for Inker and Pliny making Kim feel comfortable and reassuring her that she would not be ridiculed, this witness would never had come forward. It was also comforting to Kim that she was standing shoulder to shoulder with some of the most respected members of her community who had also witnessed Bigfoot.

Location of Sighting

My Bigfoot research in Hoopa has indicated three entry points into the Mill Creek Subdivision. One point was from the high mountains entering behind Raven Ullibarri's residence. The second point was off the highway immediately adjacent to Mill Creek. This was the street that Mary and Inker McCovey took when they saw the Bigfoot on the Highway. The third entry is the street that Kim took. This street leaves the highway and takes an immediate hard right turn and then climbs up to the top of a small hill where Paul James resides. The ride to the top of the hill is lined with berry bushes and large trees. Once at the top, the road splits and one direction goes to Upper Mill Creek Road and the other goes to Lower Mill Creek Road. This is probably the central point for all sightings in this region.

The area below this location includes the region where Debbie Carpenter saw her Bigfoot. If you follow the sightings and locations you could assume that the Bigfoot seen by Kim probably came from the hills behind Raven Ullibarri's residence. You could also surmise that the sighting made by Debbie Carpenter could have come from the exact area where Kim made her observation.

The time of this sighting is slightly early for many of the sightings in this region. It is close in time to the sighting made by Raven, but quite early for the average sighting made in Hoopa. When Kim saw Bigfoot, all of the entry points in Mill Creek were open (there had been construction in the area temporarily closing one point) and that spread vehicular traffic entering the region into two basic areas. All of the traffic coming from Hoopa (a majority of all of the traffic) would enter Mill Creek from the creek-side entrance, while traffic coming from Weitchpec (very few cars) would use the entrance where Kim saw Bigfoot. If Bigfoot didn't expect a lot of vehicular traffic, then the creature standing on the shoulder of the road while Kim made her approach is quite believable. The escape route that Kim saw the creature take would put the creature on the small hillside next to the highway.

Forensic Sketch

Harvey's sketch of Kim Peters' creature seen in the Mill Creek area.

Kim is someone who has always been afraid of the creature that she saw. From the first day that I met her until she arrived for the sketch, her fear has been consistent. She was happy to accommodate our interests in getting her sighting on paper by having Harvey draw the sketch. She came to the hotel with her daughter. Harvey's wife, Gina, kept the daughter busy drawing while Harvey questioned Kim about the sighting.

Kim first described the creature to Harvey as approximately 6–8 feet tall with a very heavy build. She stated that she thought it was dark in color, either a dark brown or black. She said that the creature did look at her vehicle and then turned away abruptly when the light hit it. She didn't get a great look but could describe the body and remembered it well. Kim also states that the creature had very shaggy hair and appeared by its build to be healthy. The last item that Kim described to Harvey was the length of the creature's arms. She stated that they were much longer than human arms, and they hung down near its knees.

Harvey presented the drawing to Kim several times for slight changes and corrections. At the end of the meeting Kim stated that the drawing was an excellent representation of what she saw on Upper Mill Creek Road.

Charles McCovey

Forestry Worker
Hoopa

Sighting #33: November 3, 2006

In the vast number of people I have met in Hoopa, few have been as reluctant to provide an interview as the tribal members of the Forestry Management Group. I have been told that many of these members have had continuous sightings throughout the valley, but several won't return calls and won't answer letters. Several people who work inside the tribal management group have told me that the Tribal Forestry Management Group is concerned about Bigfoot becoming a protected species and what that designation would bring to the organization. If Bigfoot is deemed endangered, then it is possible that the tribal forestry lands could become protected, and this may have a profound affect on their ability to cut and harvest their lands.

I was very fortunate to have Inker McCovey as a close friend. Inker convinced Charles that I was sincere and that he should grant an interview. I interviewed Charles exactly 14 days after his encounter.

During Thanksgiving week of 2006 I was at my home in the Bay Area writing notes from my previous week's trip to Hoopa. I received a phone call from an excited Inker stating that his brother had just contacted him about a Bigfoot sighting. Inker explained that his brother had never believed in Bigfoot, but he had just witnessed it.

Inker took a colleague and me to his brother's house on Lower Mill Creek Road. It was situated between Upper and Lower Mill Creek, somewhat set off from the creek itself. Charles's residence sits just below Paul James's residence. I met Charles after he'd spent a long day of working in the forests around Hoopa. He greeted us with a big smile and open arms. We talked in his workshop where he had been tinkering on some projects.

Charles explained that he was a 54-year-old Hoopa tribal member who spent his entire life in the Hoopa valley. He said that he played football, basketball and baseball for Hoopa High and graduated in 1970. After graduation he went on to the College of the Redwoods where he entered a Police Science Program to become a police officer. He explained how he

Charles McCovey after a long days work in the forests of Hoopa.

had spent his entire life hunting in the Hoopa region for bears and deer and when hunting wasn't good, going fishing for steelhead and salmon. He said that he always enjoyed the outdoors and spent time camping in the summer and spring as often as possible.

Charles said that he was assigned to work on a team that was logging an area four miles above Upper Mill Creek Road (inside Bigfoot Alley). This area is a region he knows quite well. He said that he has hunted this area a lot and has taken many deer directly in the spot they were working. He stated that it was on November 13, 2006 at 8:00 a.m. when the team arrived at the location. They worked for approximately 90 minutes, cutting, clearing and chopping wood, and decided to take a break. Charles said that he didn't know the team he was with very well and decided to take a walk by himself up the ridgeline. He told me that he wasn't wearing any cologne or hair spray and was wearing a thick coat because there was a heavy mist and fog in the air. He said that at times there was almost a very slight sprinkle in the air.

Charles walked approximately one-quarter of a mile from his crew up the ridgeline when he started to hear footsteps coming towards him, so he stopped. He stated that when he stopped, the footsteps stopped. When he started to walk again, the footsteps started again. He explained that this start and stop continued for three revolutions until he reached a high spot in the ground. He said he stood on top of a small shale formation and looked through the mist in the direction of the steps. He described the steps as not light and small like a deer, but loud and rough like a human or large bear.

From the top of his shale perch, Charles said he was peering into the mist as the steps grew louder and louder. He could finally see a figure come out of the fog and into sight. Charles said he initially thought it was someone from his work group wearing a large coat with a hood, but that thought quickly evaporated. He said he saw a huge creature, over seven feet tall walking towards him. The creature was covered with hair/fur and was walking on two feet like a human. He stated that the creature swung its arms as it walked and took huge strides. He guesses that he witnessed the creature take 10–15 strides towards him before he became very nervous and yelled at it. He said that he didn't want it to be startled by suddenly looking up and seeing him.

Charles stated that when he yelled the creature turned in several directions in an attempt to see what direction the yell was coming from. The creature couldn't turn its head to look, but had to turn its entire upper body as one unit to look in another direction. It was as though the creature didn't have a neck that could turn and allow the head to rotate.

Charles didn't get close enough to identify distinctive facial features but he could tell that it had a flat face and black shiny hair/fur over its entire body, except its face. I showed Charles the logo for California Bigfoot Search and asked him if the upper body of the creature matched the logo, and he stated that it did. He further explained that the creature's arms were very long and extended to just above the knee. The arms swung from far in front to far behind as the creature was striding toward him. He said that it did not appear that the arms were bending at the elbow as a human's arms would bend in the act of running. He described the legs as very thick and the hips as very large. He said that he felt that the shoulders on the creature were not huge and probably not bigger than the waistline. One interesting notation made by Charles was that he felt that the hair/fur covering the creature's feet was very long and the hair/fur on the lower leg almost covered the feet like an umbrella as it walked. (This observation has never been made before by a witness I've interviewed.)

Charles said that after he yelled at the creature, he thought he heard someone from his camp yell back at him. He said he turned briefly to look back at his work team and lost sight of the creature for just a few seconds; when he turned back around to look at the creature, it was gone. Charles said he couldn't believe how fast it simply disappeared. He then turned and walked directly back to the work location and said nothing to work team members, but when he came home he did tell family members.

Charles was emphatic that he never believed in Bigfoot or other people's claims. He explained that after this sighting, his beliefs of what are out in the woods have changed forever. Charles signed an affidavit.

Location of Sighting

Raven Ullibarri's residence is the closest house to this sighting. It is four miles down the road from where Charles's work crew was doing their forestry. The area where Charles made this sighting is also very close to a sighting of a Bigfoot print found 10 years ago near a spring, and aligns perfectly with all of the sightings on Upper and Lower Mill Creek Road. Charles inferred there were a large number of deer in this area, and further follow-up determined large amounts of mushrooms that also proliferate, indicating that there is an abundance of food for Bigfoot on these mountains.

This sighting was made directly on the ridgeline leading down to the Mill Creek neighborhood and in the middle of Bigfoot Alley. All of the sightings along the Mill Creek ridge and throughout Bigfoot Alley support the theory that Bigfoot is a ridgewalker.

Update

I had attempted to get Charles and our sketch artist together, but the schedules were never able to line up and Charles never met Harvey.

Final note, my colleague was with me the entire time Charles was questioned and debriefed. At the end of the two-hour session my partner told me that he felt Charles was incredibly honest and believed that he had seen a Bigfoot.

Chapter 6
Forensic Sketch Art

Harvey Pratt

Many of the skeptics I've met during my time researching Bigfoot tend to rely on the argument that says that the witness made a mistaken identification. The skeptic wants to believe that the witness saw a bear, a man dressed in an animal costume or some other easy explanation that is difficult to refute. I have heard these arguments hundreds of times and knew from the beginning how I would address them.

In my years as a police officer I was always fortunate to work with two the best sketch artists in the world. Both artists were very professional, law-enforcement oriented, and great artists. They had an ability to get inside the mind of the witness/victim and extract the portrait of the suspect and replicate it on paper. The sketches that came from the meeting between the artist and witness were always later compared to the suspect arrested for the crime. If you compared the sketch to the suspect's booking photo, it would appear that the artist drew the sketch with the suspect sitting in the room; they are that close. I always wanted to determine how it could work if you put a Bigfoot witness in the room with a professional forensic artist. Could the artist extract from the witness the Bigfoot description to the point of making the witness feel as though they "nailed" the identification? Would the effort be fruitless because the sketch artist is trained on human faces only? These questions and concerns didn't stop me from attempting to put a sketch artist and Bigfoot witness together.

The first hurdle in any hiring process, especially for a sketch artist, is to find a person who is the exact fit for the job. Hoopa is not a culturally diverse California city; it is predominantly Native American and, as such, has its own built-in biases and concerns. Hoopa is in a rural area, somewhat isolated from major cities. Some of the Hoopa residents have felt exploited in the past by others who have entered their community, taken their resources, and the left without concern for the welfare of the Hoopa people. Bigfoot witnesses in Hoopa range from 17 to 87. The witnesses are students and teachers, laborers and professionals, men and women, pessimists and optimists — a complete cross section of the society.

In reviewing the Hoopa environment and understanding the cultural aspects of the community, it became apparent that the forensic sketch artist hired needed to be someone who knew the cultural aspects of the Native American intimately, and could relate to a variety of ages and backgrounds. The artist had to have a background as a forensic sketch artist as well as experience in cultural sensitivity, not an easy bill to fill. If the person chosen had a law-enforcement background, they needed to have a low-key demeanor, be easy to understand, and a non-offensive interview style. The last, but certainly not least, important criterion was that the artist had to have an open mind about the possibility of the existence of Bigfoot.

Harvey Pratt working on a Bigfoot drawing in Hoopa.

My search for the sketch artist started with Googling "forensic sketch artist." After the search, I spent considerable time reviewing all of the applicable hits. I made over 15 phone calls and had long conversations with many of the artists; none seemed to fit. I again went back to Google and tweaked the search. One of the last sites that I reviewed showed a Native American artist who worked for a state police department. It also showed that the individual had training with the Federal Bureau of Identification (FBI). He was sketch artist Harvey Pratt.

I spoke to Harvey by phone and felt an immediate connection and a comfortable flow of communication between us. During our initial conversations I didn't advise Harvey what he would be drawing, but I questioned him about his specific artistic abilities and flexibility on travel. After those initial conversations, it was clear that Harvey was our possible artist. I later explained what he would be drawing and casually moved onto another topic. Harvey was fascinated with the topic and even confided that he had often discussed Bigfoot with his wife. He stated that Bigfoot had been spotted fairly close to their house in Oklahoma. Prior to finalizing our agreement, I asked Harvey to forward his resume, which he did. Harvey Pratt's resume looks like it was custom made for the job.

Born: El Reno, Oklahoma
Member: Cheyenne and Arapaho Tribes
Total law enforcement experience: 40 years

Education
University of Central Oklahoma
Oklahoma State University
Federal Bureau of Investigation, National Academy

Military Service
United States Marines, 1962–1965
Member, first combat unit into Vietnam

Law Enforcement Experience
1967–1972: Police Officer, Midwest Police Department, Oklahoma.
1968: Did first law enforcement drawing of a homicide suspect.
1972: Joined Oklahoma State Bureau of Investigation (OSBI).
Special Assignments: Narcotics, Special Investigations, Intelligence.
1990: Promoted to Assistant Director (OSBI).
Responsible for 400 employees, and developed automated fingerprint systems and computerized state law enforcement systems.

The only OSBI agent doing forensic sketch art and completed sketches for the following cases:
Henry Lucas/Otis O'toole
Green River Killer
BTK Killer
Ted Bundy
I-5 Killer

Murrah Federal Building Bombing
Randolph Dial
Many other homicides and violent crime cases were also worked to conclusion via sketches.

Specific forensic art expertise
Witness Descriptive Drawing
Skull Tracing
Age Progression
Soft Tissue Reconstruction
Restoration of Photographs

Special recognition
Inductee, Southern Cheyenne Chief's Lodge
Recognized as one of the traditional Cheyenne Peace Chiefs
Recognized by Cheyenne people as "Outstanding Southern Cheyenne"
Recognized as a Master Native American Artist
Completed art projects in oil, acrylic, watercolor, metal, clay and wood

---

When I received the resume I couldn't believe my eyes. Harvey appeared to be born, raised, trained, educated and professionally molded specifically for working on the Hoopa Bigfoot Research project. Nobody else that I interviewed came close to his credentials. When I spoke to Harvey and confirmed with him that we had agreed on his price and specific job issues, he was elated. He told me, "My medicine is strong on this issue."

On March 19, 2007 Harvey and his wife, Gina, arrived in Hoopa. They met me in my room and there was immediate chemistry among all of us. Harvey and his wife are very warm, caring people who were as interested in the Bigfoot project as I was in listening about their past experiences. I gave Harvey and Gina the agenda for the week and explained that I would do everything possible to get Harvey out of his meetings and into the mountains at every opportunity possible. I had scheduled four interviews per day at two-and-a-half-hour intervals, with accommodations for meals and breaks. It was a full schedule.

Interview Day 1

I had attempted to mix the witnesses by age, date of sighting and sex so that Harvey was not deluged in any one day by a consistent stream of the same type of sightings. At the end of each interview we would thank the witnesses, discuss their drawing, answer questions and then de-brief after they left. The first day the interviews were in the following order:

Damon Colegrove
Juliene McCovey
Romeo McCovey
Raven Ullibarri

I must admit that after Damon's drawing I was slightly stunned. In my years of researching Bigfoot I had heard from many witnesses that Bigfoot was a mix between human and ape or gorilla. I had read many accounts of hunters who were reluctant or unwilling to shoot at Bigfoot because it looked too human. The only Bigfoot facial pictorial that existed on film was the Bluff Creek video. The face on that footage was covered with hair and looked somewhat similar to a gorilla with some human characteristics. When Harvey had completed Colegrove's drawing my thoughts about Bigfoot changed forever.

Damon was very sure about what he witnessed. Damon is an artist and thus looks for finite detail. The detail that Damon recalled helped immensely in Harvey's drawing. The most interesting part was the facial structure that Harvey was able to re-create; it looked very much a cross between human and animal. If Damon had not been sure about the accuracy of Harvey's drawing I would question it, but Damon was positive that Harvey had nailed it.

The second witness originally scheduled had to cancel. I had wanted to talk with Juliene in the previous week, but she was out of town. The Tuesday morning that the interviews started, I saw Juliene in front of the market and asked if she could participate that morning. Being the flexible person that she is, Juliene accepted. This was an important interview because Juliene had stated from the beginning that she felt her Bigfoot was female, one of the few witnesses who made that claim.

Colegrove Drawing

Juliene Drawing

Harvey worked well with Juliene and was able to produce a drawing that reflected the subtleties of a female. Juliene explained to Harvey that she felt the creature was pregnant, had another smaller creature on its stomach or had a very large abdomen. She also believed that the creature was smiling at her as she walked past.

Juliene's drawing had more characteristics of an animal than human, but it was close. You could see a human quality about the drawing, and Harvey was able to draw it to Juliene's satisfaction. I should state that all of the witnesses were willing to tell Harvey when he did not have the drawing correct. They didn't accept first, second and sometimes third drawing renditions. There was always a good dialogue between Harvey and the witness, and Harvey never showed indications of being defensive of his drawings. Note that Juliene's creature has more jutting jaws but it also has the long hair that appears in Damon's sketch.

The first afternoon started with Romeo McCovey. I knew that Romeo had seen the creature at night without any direct lighting, thus making the sketch a challenge. Romeo made a great effort, with Harvey's assistance, but couldn't add significant detail, other than what appears in the drawing. I appreciated Romeo's honesty and straightforward approach to the draw-

Romeo's sketch — Raven Facial Sketch — Raven Back Sketch

ing as he could have easily fabricated many features, which he didn't. The drawing didn't take long to complete and Romeo appeared to feel bad that it didn't have more detail. We did glean some valuable information from the drawing: the massive shoulders, hairy facial area and how tall the creature was when it stood up. The location of Romeo's sighting was also significant as it occurred very near sightings by Inker and Kim Peterson.

The person who had the longest and most intense Bigfoot experience was the final interview for Tuesday, Raven Ullibarri. Raven had seen the Bigfoot who visited her backyard, got a great close-up look and then had police officers respond to the trespassing and prowling. Raven was a superb witness who obviously has that memory permanently etched into her brain. Raven was able to stay poised enough to take a good look at the creature's face, watch it leave her property, and then have the ability to stand and watch. Remarkable! We knew that Raven had one of the best looks at a Bigfoot, but we didn't know she would be able to provide guidance for two drawings of different views. I do believe that the profile sketch that Harvey completed for Raven is one of the best of the week. Raven was able to detail specific aspects of the face that assisted in bringing out personality and character. It's interesting to note that this creature also had flowing hair down the back, a large forehead and facial hair. It's also important to note the length of the creature's arms relative to the length of the body. Raven insisted that the arms hung near the creature's knees. She also described how the creature was carrying a large garbage bag in one hand and the bag didn't touch the ground; that's how tall it was. The creature's eyes have a definite human quality and don't have the jutting and hooded brows that you see with primates. The mouth area does have a slight bulb to it like a primate, but the general facial structure is soft enough to show human qualities.

There is not a great distance between Romeo's sighting and Raven's sighting, maybe half mile. From Raven's house to Kim Peterson's sighting is less than a quarter mile and less than one-eighth of a mile to the Pliny McCovey sighting. It could be possible that the creature observed by Romeo is the same as seen by Raven.

At the end of the first day of sketches, Harvey, Gina and I sat in the room and were mesmerized. The drawings were very revealing and not what we were expecting. The Pratts made the comment that they felt the witnesses were very credible and surely saw what they claimed. Harvey has 40 years of law enforcement experience, Gina 15 and myself 30. There was, therefore, a combined 85 years of law-enforcement experience judging the credibility and integrity of each witness. The first day's verdict was in; all of the witnesses were incredible, and all did a great job describing their experience and their Bigfoot.

Interview Day 2

The second day gave Harvey an interesting mix of witnesses. Two witness sightings were made from a car and one from a bed. Two witnesses were tribal members and one was a store manager. Two of the witnesses were female and one male, and the ages were spread quite far apart.

Day two started with Tane Pai Wik. Tane saw Bigfoot when she was a little girl lying in bed with her aunt. Tane said that the memory of seeing Bigfoot was burned into her brain and she will never forget the experience. Tane's husband, Corky, dropped her off at the hotel. She quickly became engaged with Harvey on the project and the sketch came together rapidly. Tane had done a drawing of the creature in an art class in high school and even allowed me to photograph the drawing when I was at her residence. There are similarities between what Tane drew in high school and the sketch that Harvey produced. Harvey was not allowed to view Tane's pic-

Harvey's sketch for Tane.

Tane's drawing in high school.

ture until he was completely finished with the sketch. At the conclusion of the meeting, Tane expressed her appreciation with what Harvey produced and felt it was a great rendition of the Bigfoot she witnessed.

Corky arrived at the hotel and picked up Tane when she was completed with her sketch. Gina Harvey and I had a post-sketch meeting. It was our general feeling that the sketch was accurate. Tane had made a point of saying that this wasn't a human; it was a creature. When Corky came into the room to pick Tane up, he looked at the sketch and stated that the picture was of a lost mystical Indian. Tane felt was that it wasn't a Native American, but it appeared to her to be half human and half creature. She explicitly mentioned that the creature's teeth were green and it had a hairy upper body, head and a face that was not human. The reality of Harvey and Tane's sketch is that there is a human factor in the face of each drawing. Tane had told me in earlier conversations that she felt this creature was old and it might have actually lived in the cabin where they were sleeping. This was an idea that she and her aunt had discussed after the sighting. You can see that Harvey's drawing shows an aged quality to the sketch, and the creature does appear to have less hair on the top of its head compared to other sketches completed that week.

Mularkey's sketch.

Martin sketch.

The second interview on the second day was with the most pessimistic witness prior to his sighting that I'd ever met. Michael Mularkey made it a point to tell me that he'd heard all about Bigfoot in his time in Hoopa, and always believed that it was an old Native American legend that held no truth. Michael wanted to make it clear that he wasn't looking to see Bigfoot because he knew it wasn't real. He is a believer now. Michael is so soft spoken that he flies under the radar in Hoopa. He doesn't talk to the general public about his sighting; you almost have to pry it out of him.

Michael met Harvey and appeared to be a bit skeptical of the entire process. He had some serious personal issues that were occurring during the interview and we offered him the opportunity to come back later; he refused. As the interview progressed and the sketch began to come together, he started to concentrate even more. Michael was a witness who had a good look at the creature, but not a close-up view. The creature was caught in Michael's headlights as he saw it walk across the road in front of his vehicle. Michael was able to provide minimal facial information. He made a point of stating that the creature's arms and legs were long, and that it was generally huge. The sketch possessed the same long hair flowing down the back that previous witnesses had described.

Michael's drawing was one of the first completed of the entire body of a creature. When the drawing was done we didn't think a lot about the body structure or have great credence in Michael's description until more of the drawings were completed and they all started to appear similar. Michael's enthusiasm and participation continued to improve as the sketch came together. When Harvey was finally done, Michael couldn't believe how he was able to accomplish what he did. Michael was amazed.

The third and last witness of the day was Jackie Martins. Jackie had witnessed the creature when she was 17 and driving into Hoopa from Bald Hills. Jackie and a friend were in a vehicle when the creature crossed in front of their car and went down a hillside. Jackie was very excited about working on the sketch project. She was someone who Harvey and Gina wanted to meet because of her Hoopa language background and her former position as a council member for the tribe.

Jackie's drawing had some very interesting aspects to it that matched with other drawings. The creature had very long arms, had a huge chest and shoulders and essentially no neck. If you look at all of the drawings to this point, there is a consistent theme that doesn't match with the vast majority of current Bigfoot drawings. Many of the drawings that you'll see on the Internet have a significant level of ape/gorilla appearance. The sketches completed to this point in Hoopa have a significant human quality to them.

Michael and Jackie's drawings are very similar in body appearance and stature. The sightings were made approximately 40 miles from each other and on different sides of the reservation. Michael's was at the south end and Jackie's was on Bald Hills Road that skirts the northwest corner.

At the completion of the second day, Harvey and Gina again commented on the sincerity and honesty of the witnesses interviewed. Both felt that the witnesses were credible and honest, and they felt the entire experience of sketching the Bigfoot and talking to the witnesses about their sightings was exhilarating. We were all impressed with the size of the creature that each person described, and the similarities in the physical attributes of each creature.

Interview Day 3

This was going to be Harvey's busiest day, for many reasons. Harvey had four interviews scheduled and we had to take a road trip of 30 miles to meet one of the witnesses.

The first witness on the third day was Ed Masten. Over the last 24 months Ed and I had become good friends and I'd seen him on almost every trip to Hoopa. Ed is someone who would do anything for you and would go out of his way to assist in some way. As in past interviews, I did not tell Harvey or Gina anything about the individual except the specifics about their role on the reservation and their sighting. Ed was very punctual for his meeting and was eager to start the process.

You could see that Ed had a specific idea of the creature etched in his mind. The creature had obviously left an eternal image on Ed's psyche because he had no problem describing and eventually changing details on the sketch. Ed and Harvey meandered their way through the process with a series of changes and explanations all smoothly accomplished.

Ed explained how he had encountered the creature twice in the Tish Tang basin. The best look he had of the creature was when it was standing partially behind a tree. As Harvey started to draw what Ed observed, you could tell that this drawing was going to be much different than the first seven. When the drawing was finished, it looked very much like the creature depicted in the famous Patterson Gimlin Bluff Creek video. The creature had hair covering its entire face, had a look somewhat of a gorilla or ape, yet he stated it ran more like a human. The creature did have similarities to other drawings in that it didn't appear to have a neck. In Ed's description, the creature had hair lightly covering its forehead. There was also hair covering almost every region of the face, different than other witness descriptions in other sightings. The face was also more round with less of an abrupt jawline than seen in other drawings. There is the distinct lack of a high, hairless forehead in Ed's drawing. When Harvey was completed, Ed was very happy with the drawing and stated that it was a great depiction of what he saw in the woods.

Witness number two, on day three, was Josephine Peters. Josephine was the oldest of the witness group, and was also one of the few that has had multiple encounters with a Bigfoot. Josephine was over 80 years old

Ed Masten sketch. Josephine Peters sketch.

but had a very sharp mind and positive attitude about this entire process. She was at the point in life where it was very difficult for her to walk or drive, so her friend brought her to the hotel. We sat Josephine in the room and had a great conversation about Hoopa, lifestyles in the early 1900s and what life was like today. Josephine readily talked about stories that her mother and grandmother had told her about life in Hoopa in the late 1800s and how the valley had changed.

Harvey got Josephine to the point of starting to describe the creature and its physical appearance as she witnessed it at her fence. She explained how the creature had come down from the creek near her backyard and that she thinks that it must not feel threatened and that's why it has visited. She stated that she had a very good look at the facial area of the creature and she felt that she could assist in that description.

Harvey worked with Josephine on just the facial area of the Bigfoot. There were specific people we were targeting for specific areas of the body to draw, and Josephine was someone we had hoped could provide an excellent facial description. She didn't let us down. After 30 minutes of describing the face, you could see the personality of the creature start to shine through on Harvey's drawing. There were minor changes that Harvey made as he progressed, but Josephine was doing an excellent job answering his questions and working with him on the colors and shading.

Josephine's sketch resembled several others that Harvey had completed earlier in the week. It had a high hairless (or nearly hairless) forehead, a light beard like a human male, long hair flowing off the back of the head, and eyebrows that are not quite as heavy as an animal. Josephine explained the creature did not appear to have a neck, and she felt that the sketch depicted it adequately.

After almost two hours of sketching, refining and changing, Josephine told Harvey that he had it perfect.

After Josephine had left the room Gina, Harvey and I had a long talk about the differences between the Ed Masten sketch and Josephine's. Each witness was believable and credible, yet both sketches were very different.

After the first two interviews of this day Harvey, Gina and I made a road trip to Orleans to interview our third witness, Mary McClelland. During the ride to Orleans the Pratts were able to see the countryside, Bluff Creek, the Klamath River and understand the landscape that Bigfoot roamed. They were fascinated by how rugged the mountains and forests were, and how thick the trees and brush were in and around the rivers. There were several places that we stopped and took photos, as Harvey had stated that they didn't have hills and rivers like this in Oklahoma. During our trip we talked about the first two sketches of the day, and were fascinated at the differences in the creatures and the sincerity and credibility of the witnesses.

We arrived in Orleans where we had agreed to meet Mary at the lone restaurant in town. The owners were extremely hospitable and allowed us to sit in the back area of the bar away from people. The location gave us the solitude that we needed and the lighting that was ideal for Harvey. Mary first apologized for making us drive to Orleans, but stated that she was babysitting her boyfriend's kids and it was difficult for her to get away. She thanked us for making the trip and said she was eager to get started.

Mary began talking to Harvey about the position that she initially saw the Bigfoot and the best angle she had of seeing it. They both agreed that Harvey would draw a sketch of the creature in a slight crouch, getting up from the ground and looking back slightly. As the drawing started to take shape you could start to see its enormous size. The proportions were not human, but its face was human-like, as in the other drawings. Mary made a point of stating that she didn't see the face clearly enough to complete

Mary McClelland sketch.

Kim Peters sketch.

a facial profile sketch, but she did see it a little. She described that the creature had very long arms, much longer than a human's. It had huge shoulders and legs, much larger than any human's. You could actually see Mary start to smile as Harvey came to near completion of the drawing. She was amazed at Harvey's ability to take her words and place them on paper in the figure she was attempting to describe.

Mary was with Inker McCovey when they saw the Bigfoot. She was excited to see how Inker would interpret her sketch and how accurate her sketch would be to Inker's recollection.

At the conclusion of Mary's interview, we stayed at the restaurant and had dinner. I encouraged Gina to have a piece of their homemade cheesecake; she agreed it's great. We made the early evening drive back to Hoopa for interview number four. On the ride back Harvey commented on how thorough Mary was with her descriptions and how Harvey felt that the drawing was accurate. He felt she was very credible. Mary's drawing also showed the same facial beard that is present in many of the drawings, hair flowing towards the shoulders, massive upper legs and shoulders.

Kim Peters was the last sketch of the day. Kim arrived after working a long day at her pre-school. Kim was always someone who had a kind word and a cheerful personality. She entered the hotel room and I introduced Harvey and Gina. One of the first things that Kim told Harvey was that she was a person who never believed that this creature existed — ever. She had made it clear to people in the community that she would be respectful of their Bigfoot statements, but held her own belief that this was either an illusion, dream or another animal that looked like a Bigfoot. Kim now states that her mind has totally changed since seeing the creature, and now she is positive that Bigfoot does exist. Kim and Michael Mularkey had similar beliefs before seeing the creature and now both are true believers.

Kim's drawing started to come together and looked very similar to Mary McClelland's drawing. The similarities in the descriptions made some sense since both sightings were made less than a quarter mile from each other. As far as I know, Kim does not know Mary, and Kim had no idea that we had met with Mary earlier in the day. It is an absolute fact that Kim never saw Mary's sketch.

Kim's description included a statement that the creature had a large chest, huge shoulders and arms that extended to near the knees. She made a special point of stating that the legs looked very powerful. She said that the creature's entire upper body turned as it looked in her direction and it appeared that the creature didn't have a neck like a human. Kim did the best she could on the facial description but with the length of the sighting and the lighting conditions, the facial description did not have significant definition. Kim did say that the creature had hair on almost its entire body.

At the end of the meeting with Harvey, Kim made the same statement to him that she made to me when I first interviewed her. Kim said that the sighting was frightening to her and she went straight home, locked the doors and didn't go outside at night for many weeks. At the session's conclusion Harvey showed Kim the finished product; she was very happy. Kim confirmed that the drawing represented what she saw and she thanked Harvey and myself for the opportunity to sketch the creature.

At the close the evening Harvey, Gina and I talked about the day's sketches and went back to discuss the similarities between Kim's drawing and Mary's drawing. They were very close. I reminded Harvey that the locations of the sightings were close in proximity and it may be possible to attribute some of the similarities to this; maybe they are the same creature. Harvey said that the descriptions and the drawings were too close not to acknowledge their similarities. He couldn't say positively that both witnesses saw the same creature, but they were close. We also talked about Ed Masten's sketch at length. Harvey felt that Ed was being honest about what he saw, but we were still somewhat surprised at the differences between Ed's sketch and the sketches of the other witnesses. Ed's creature had the most facial characteristics of an animal (gorilla/ape) while the other sketches had facial features that could be attributed to being human. Harvey felt that Ed was a very sensitive person who cared about doing a great job on the sketch and spent a lot of time explaining specific issues that he thought weren't correct. We all felt that Ed would not make a mistake on the drawing and the result is what he exactly observed.

Interview Day 4

Day four of sketches started with the one person in the Hoopa tribe that I'd spent the most time communicating and meeting with, and the one person in the tribe that had the most encounters with Bigfoot, Inker McCovey. Inker had told me several times that he was looking forward to working on the sketches, and he was interested in learning more about Harvey. Inker was one of the few actually early for his appointment.

I made the introductions and Inker immediately took his seat and started to tell Harvey about the three encounters he had with Bigfoot.

I had already briefed Harvey about Inker's encounters and he knew that "Ink" was the driver of the vehicle Mary McClelland was in when they saw a Bigfoot. Harvey had decided to avoid having Inker complete a second sketch on the Mary/Inker incident and had him focus on the arm that entered his bathroom as a child and the two Bigfoot he had observed on the hillside.

The first sketch drawn for Inker was the arm in the bathroom. Inker explained to Harvey how he was a little boy sitting on the toilet when the Bigfoot arm reached inside an open window and attempted to grab him. Inker was questioned about the possible motive of the Bigfoot, and we briefly discussed it as a group. Inker did confirm that there was always a small garbage can in the bathroom adjacent to the toilet, but in the general direction of where the arm was reaching. Inker did confirm that there were females living in the house. Could Bigfoot have been reaching for a sanitary napkin in the garbage, or maybe even made a mistake and thought Inker was a female? If we review the issues surrounding Raven's sighting and the sanitary napkins in her garbage, maybe the same issues involve Inker. I did ask every elder I had met in the Hoopa tribe if they had ever heard of a boy being kidnapped by a Bigfoot, and the answers were a unanimous, "NO."

Harvey asked direct, pertinent questions about the Bigfoot arm, size, etc. Inker remembered that the Bigfoot reached in until the bicep completely filled up the window frame (24 x 36 inches). It was huge! Inker said that the arm was very large compared to a normal human arm. He also stated that he remembered that the thumb on the creature didn't look or work the same as a human's. He stated that he thought the thumb worked like the thumb of an ape, it would grasp in the same direction as the fingers. He stated that he distinctly remembered that the thumb of the creature had a very dark nail but couldn't remember if the fingers had nails. Inker recounted how the outside portion of the creature's arm had hair that was approximately four inches long but the hair wasn't thick; the inside of the arm didn't have any hair.

Inker told Harvey he had screamed and squirmed his way off the toilet, and how his mother had run in and found him on the floor. He stated that his dad grabbed the rifle and flashlight and chased the creature down into Mill Creek, and eventually lost it. He lived in a very old house, long since demolished, on the property where he still resides. This is the same property that is still in the middle of sightings by an entire community 45 years later; it's also in the middle of Bigfoot Alley.

Inker's property sits about 400 feet from the start of Lower Mill Creek Road and contains approximately 300 feet of Mill Creek frontage. Mill Creek is directly behind their residence and this is the area where the creature escaped in Inker's bathroom scare; it is also the location of the escape in the Hank Masten incident. The area directly in front of Inker's house is Lower Mill Creek Road and across that street is a steep hillside where there are no homes, but it does contain heavy brush and trees. If a creature wanted to watch on this neighborhood, Lower and Upper Mill Creek Road would be optimum locations. If you drew a straight line from

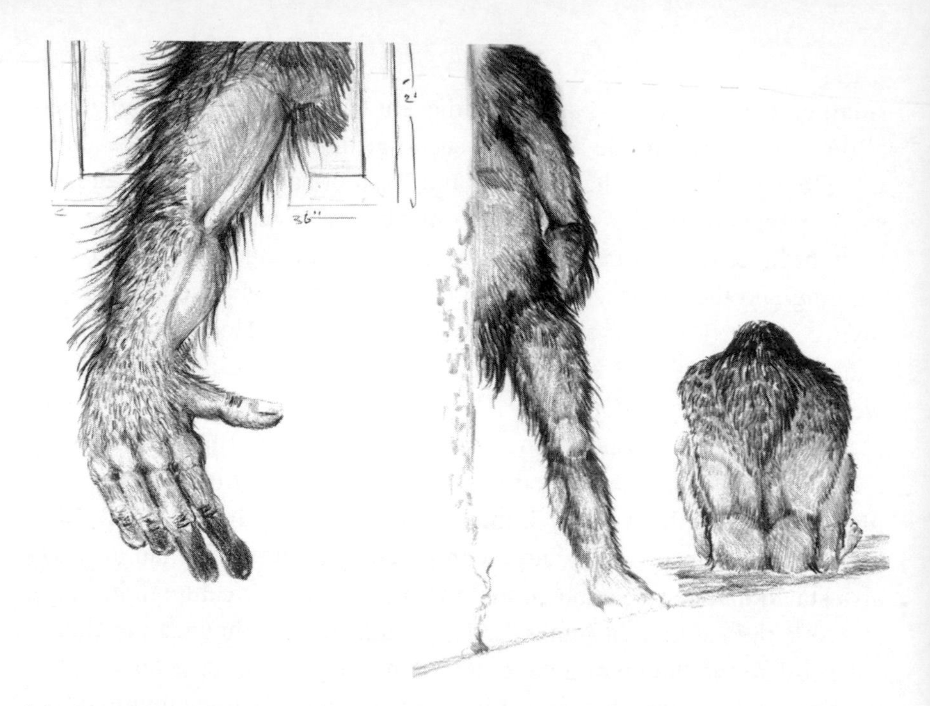

Inker's creature's arm.

Inker's hillside sighting sketch.

Raven Ullibarri's residence on Upper Mill Creek Road to Inker's residence on Lower Mill Creek Road, it would cover approximately 300 feet. On the night of Raven's incident it is possible that the creature ran down the hill, across Inker's yard and into the creek, but this couldn't have happened until the officers had cleared the scene.

There is an identifiable 45-year history of sightings on Mill Creek Road. Something is bringing the creatures back to this neighborhood looking for human interaction, or accidentally having human contact. The sightings on the road appear to be accidental, but we can't definitively state that. The motives of the creature reaching into Inker's bathroom could be as innocent as wanting human contact, a touch, or something even more sinister, we don't know.

Inker's second sketch was his sighting in the hills above his residence. Inker explained how he was hiking and thought he saw a stump on the opposite hillside. He said it caught his eye because he thought he saw movement. Inker sat and watched the figure crouch into a small ball and then the larger creature partially concealed itself behind a large fir tree. It appeared to Inker that the larger creature was the parent and the smaller creature was the child. He made it clear that if he hadn't seen the creature move that was in the ball, he would have walked by and never noticed it. He said that after he watched it for a few seconds he knew he was looking at two Bigfoot.

The behavior of these creatures fits with many animals in the wild. A deer will freeze in the brush and allow a hunter to walk by before it eventually moves. A mountain lion will sit in a tree and watch as prey or people walk by below. What is different about this sighting is that Inker felt that the smaller creature was getting specific directions from the parent. With concealment and safety less than five feet from the smaller creature, this appears to be almost a safety and training session for the smaller Bigfoot.

In any environment where there is reproduction of the species, it is generally considered by science to be a healthy region. The observations made by witnesses in this book confirm that Hoopa is an area where Bigfoot is bringing its young and adolescent age children. These sightings indicate that Hoopa is still a preferred area for Bigfoot to venture.

The last item we discussed with Inker was the sighting he had made with Mary McClelland. Harvey had made a determination that Inker had spent significant time explaining his sightings and it would probably be prudent to show Mary's drawing to Inker and gauge his response.

Inker was shown Mary McClelland's drawing and he immediately stated that it was excellent. He pointed to the quadriceps muscle and said that it was large, as in the drawing. He said that the creature itself was very large and impressive. The pose that Mary had described was perfect for what they had seen. He stated that he could not do a better job of describing what they saw. Inker did talk briefly about he and Mary watching the creature leave the area. It appeared odd to me that the creature was facing west towards the river where there was nothing between the road and the water, yet it changed its direction in the middle of the road (costing time and space), running across a small field and then leaping a fence into a person's yard. It would appear that the creature chose to go to a yard on the mountainside of the roadway versus going straight to the river.

The last sketch of the day, and the last of the week, was with Michelle McCardie. Michelle had been in the hospital earlier in the week and was recently released. I had been in contact with her and her daughter during the week and each had encouraged me to continue to call until she was home. I eventually contacted Michelle on Friday morning and she

Michelle McCardie sketches of the creature.

advised that she did want to do the sketch after she took a morning nap. We agreed that I would take Harvey to her residence and they would conduct the interview in her living room.

Harvey, Gina and I went to Michelle's home and she was happy to see us. She was very interested to understand how the sketch process worked because it had been on her mind for so many years. Harvey immediately went to work in an effort to speed the process for Michelle's comfort.

It became apparent early in the meeting that Michelle's recollection of the incident and her ability to explain what she saw was sufficient for Harvey to complete two sketches. One sketch would be of the body and another of the face. Michelle's sighting is very significant as she and her daughter are the only people I am aware of who have seen a young Bigfoot up close, and sufficient enough to complete a forensic drawing of the creature. It was also important for researchers to understand the rate of physical development of the creature, understand the muscle composition and the associated body structure.

Harvey stated that Michelle did a very good job concentrating and explaining what she observed. She stated that the creature had a very sweet face and its body was built like that of a twelve-year-old boy. She explained to Harvey that the creature didn't show any fear, but it appeared to know that it had been caught eating berries and it was best to get out of the area. She said that it had definite intent on leaving; it didn't run, but did move swiftly. Michelle also explained how the creature was taking the berries with what appeared to be its fingertips. It was not stripping the vines in an aggressive or destructive manner, and it looked to her that its fingers had dexterity enough to take one berry at a time. It's important to remember that this creature was witnessed by a yard full of kids and they were not the catalyst for the creature leaving the area. The kids were in the yard first and the creature walked up and started to eat berries in their presence.

Harvey completed the sketches and then left the residence after approximately 90 minutes. He went back to the hotel room where he added artistic touches that Michelle explained needed to be completed to enhance the appearance of the creature. Harvey said that he was excited to have completed a sketch of the creature as an adolescent, and he felt that he had captured what Michelle had seen.

The body structure of Michelle's drawing is of a slightly built creature that does not appear to have a neck. She described the facial area as having some reddish tone, such as a younger human face may possess. She said that the creature looked friendly and imparted no fear. It did have some light beard on the chin and heavier hair length around the neck area.

Analysis

In all of the Bigfoot/Sasquatch and Yeti books, documents, studies and writings I have reviewed, I have never found a case study where the witness was interviewed in person and at length and then legally locked into their statement by an affidavit. Never. I have never found a case where a witness was interviewed to the extent detailed in this investigation and then placed in front of an FBI-recognized expert forensic artist and asked to assist in drawing what they observed. Never. What we have done in these instances is to go to the farthest investigative extremes in an effort to document the creature as it was actually observed. If we were able to capture any of these creatures, I have no doubt they would appear exact-

Damon sketch.

Tane sketch.

Raven sketch.

ly as Harvey has drawn them. Harvey Pratt's track record of sketching suspects, and the accuracy of his sketches, is too impeccably accurate to question his ability. His skill as a professional artist in drawing an assortment of animals and figures also goes to his expertise. He is an award-winning artist in his pursuits away from law enforcement.

It would appear from the drawings that some very controversial conclusions are going to be drawn. One of the first realizations is that Bigfoot appears facially to be much more human than originally thought. It has always been believed and said that Bigfoot walks on two feet like a human. Agreed. The facial features found in Bigfoot in many of Harvey's drawings could have evolved from a variety of sources. If you refer back to the Josephine Peters' sighting, she states that the Kurok and Hoopa people often escaped into the woods to avoid battles and conflicts. In many of those instances, families lost track of relatives and friends. Would it have been possible that an adult died from injury and Bigfoot came to a child's aid, raised the kids and eventually bred with them? I know that this seems a far-fetched conclusion, but based on the facial features of many of the drawings there appears to be some human DNA in the mix. There have been several hunters I have interviewed over the years who claim to have seen Bigfoot while hunting. They felt that the creature looked too human-like to shoot. This statement falls in line with the drawings from Harvey. Refer to BFRO reports on pages 325 to 327 for more of these details.

In comparing the facial structures of the three sketches by Damon, Tane and Raven, it would appear that the Damon sketch has the features

Ed Masten sketch. Josephine Peters sketch.

that most closely resemble a human. The eyebrows are not jutting, the jaw line is subtle and the nose is not as large and flattened. The same descriptors could also be used for the Tane sketch. Tane indicated that her subject appeared older, and the receding hairline would follow that description. The Raven sketch shows a creature that has a broader flattened nose, a protruding mouth and jawline, as well as eyes that appear more inset to the skull. One consistency among all of the three sketches is the flowing hair coming off the back of the head. It's interesting that none of the descriptions show the hair flowing into the eyes; it always flows to the back. Since these creatures do live in a harsh environment, maybe the hair offers some level of insulation and warmth over their shoulders and back of the head. The other obvious consistency among all three is the lack of a definitive neck. None appear to have a neck, and all witnesses describe a creature that turns its body to look and doesn't just turn its neck. It also cannot be overlooked that the complexion of these creatures is much more light than dark. There is not one creature that has been drawn that has the black complexion of an ape or gorilla.

In reviewing the sketches of Ed Masten and Josephine Peters you can readily see major differences from the Damon, Tane and Raven sketches. Harvey, Gina and myself talked at length in regard to the Ed Masten sketch. Ed's is the only sketch that has hair covering almost all areas of

Gorilla.

Masten sketch.

the face, yet the face has soft touches. It is the one sketch out of all that were completed that is extremely close in appearance to the creature caught on the Patterson–Gimlin film in Bluff Creek. The similarities between the two faces make them almost appear identical. After discussing the sketch, Harvey and Gina did believe that Ed saw a creature that matched that description. Ed is one of the few people in the tribe with whom I have personally spent a significant amount of time, and I believe in what he has said. The real question is, what does this drawing tell us?

After thinking through the sketch, the location (Tish Tang Basin) on the far southeastern edge of the reservation, we came to some conclusions. Thinking about the human race, we know that there are quite distinct differences between people. People from African nations look extremely different from Mongolians. Caucasians in Ireland look much different than Hispanics in Central Mexico. There are significant differences in primates. There are the great apes, huge and powerful, and the orangutans, smaller and less powerful. There are also smaller and more fragile monkeys from different parts of the globe. All of these primates fall into one classification, yet are very different in their physical appearance.

The lack of a neck on Bigfoot may be an evolutionary process of pure need. When a mountain lion attacks a human most of the bites occur either on the neck or at the neck. The neck is a very vulnerable area on a human, as it holds the spinal cord and carotid artery. If the cord is slightly injured, paralysis can occur. If paralysis occurs during an attack, you are dead. If the carotid artery is punctured, you can bleed to death in minutes. If humans had no neck, it would make the job of predators more difficult by making the spinal cord and arteries harder to injure.

A young Bigfoot would be vulnerable for a to predator attack; when it reaches full size there is no chance it would fall to a mountain lion or bear. A mountain lion may drag down a young and small Bigfoot, but with no neck, killing the Bigfoot is a very difficult task. I can't imagine a mountain lion attacking a large Bigfoot by grabbing it around its head; the head is too big. Since there is no neck to grab on a Bigfoot, predators would have to attack another area of the body and that would put them in jeopardy of retaliation.

There are physical appearance differences in the human race, so why wouldn't there be in the Bigfoot race? Maybe Ed's creature came from a region hundreds of miles away in a different wilderness area where facial hair features are the norm. Maybe it came from the Bluff Creek Region where Patterson and Gimlin got their footage.

The Josephine Peters sketch is almost a cross section between the Masten sketch and the sketches that depict a more human form of the creature. Josephine described a Bigfoot that has a very well defined mouth area that protrudes like a primate. It also possesses a distinctive flat nose and protruding eyebrows, all similar to primates. It is probable that the Peters Bigfoot came from the coastal range west of Hoopa. If you refer to the interview with the Hoopa Tribal Wilderness Team, and specifically Elaine Creel, you can remember she stated that they had completed genetic testing on bears and had found that the bears do not cross the Trinity River. Can we use this data and extrapolate that the Bigfoot in the region also don't cross the river to breed? I don't think so. There are too many stories where residents hear yells and screams (Bluff Creek Resort) from different sides of the Klamath and the screamers appear to eventually find each other and merge for a meeting. I think the physical differences among Bigfoot reflect different genetic groups.

In comparing a primate photo with the Ed Masten sketch I do not believe that there could be a mistake in the identity. The gorilla has an extremely dark complexion, protruding mouth and nose with deeply inset eyes. The Masten sketch doesn't have any of the characteristics as the gorilla, and it specifically has hair over the entire face, which the gorilla doesn't. The shapes of the heads are quite different, with the gorilla having almost an oblonged structure and the Masten sketch is nearly round. There is an interesting comparison between some of the sketches and the gorilla hairline. The gorilla has a hairline that starts very high in the forehead and flows to the back of the head and shoulders. Sound familiar? The physical shape of the Masten sketch has a very close similarity to human.

In reviewing several photos of gorillas, they appear to have a more definitive neck than almost all of Harvey's sketches have depicted in Bigfoot. Almost every aspect of the body structure of the gorilla is much smaller than Bigfoot. The gorilla sometimes walks on all fours; Bigfoot never does. The gorilla walks with a pronounced hunched over stature; Bigfoot doesn't. Bigfoot has been described as slightly hunched but not nearly as angled as the gorilla.

The complexion of Bigfoot is almost uniform in every description. I have never heard of a Bigfoot described as having black skin. Almost every sighting report I have taken describes a tan or brown complexion, but never white or black. Colors of the hair vary greatly. I have heard of Bigfoot with completely grey or white hair; brown and black are common, along with a tint of red or cinnamon. I have never heard of a platinum blond. Well, maybe if there is a sighting in Beverly Hills.

I included the photo of a gorilla sitting because it shows long flowing hair coming from the forearm. This is interesting because that is how Inker described the arm that entered his bathroom. The photo also shows longer hair near the shoulders, another Bigfoot trait.

Where does this lead us?

The common belief between everyday citizens and some Bigfoot researchers is that we are dealing with a hybrid of a gorilla or ape. There are many, many drawings of these creatures on the web. I have reviewed some sketches made by artists who claim to have made a forensic sketch based on a phone conversation with the witness. None of these drawings look similar to the drawings made by Harvey. The drawings made from a phone conversation have widespread presence over the web and have a definite influence over some portion of the Bigfoot research community. There is a specific reason that forensic sketches by law enforcement are usually done in person. The interaction, instant feedback and ability to judge response are a huge factor in the forensic sketch. There is a major cost associated with bringing a seasoned and well-qualified forensic artist to a Bigfoot witness, but I think the results have justified the expense.

It would appear from Harvey's sketches that Bigfoot in the Northern California and Hoopa area are much closer to a human than most tend to believe.

Gorilla.

One of the most time-consuming endeavors that I undertook in relation to this study was a periodical search. I spent countless hours rummaging through old microfiche attempting to find documents and articles that assisted in explaining the sightings in Hoopa. I was also searching for documents that specifically made reference to the physical human quality in Bigfoot in this area. I think I found the article.

It was in the late 1950s that roads were being built into the Bluff Creek area and it was also during these times that huge tracks started to appear on those roads. Subsequent to the tracks being found, locals started and coined the name "Bigfoot." Newspapers in the area ran stories about road equipment being thrown around near Bluff Creek, and concern started to rise about what was happening in those hills. Stories started to proliferate about a huge seven-to-eight-foot-tall creature that was roaming the lands, and how some of the construction crews were quitting their jobs out of fear.

Just outside of Willow Creek is the small town of Salyer. There isn't much there now, except a small store and deli and a large California Department of Forestry Fire Team. In the late 1950s Salyer had a California Conservation Corps (CCC) camp on the outskirts of town. The CCC crews build trails, work minor fires and generally are a group of older boys and young men that work hard and long hours during the summer months.

The community was concerned enough about the Bigfoot presence that they ran a front page story talking about a large Native boy who ran away from a CCC camp. The association to Bigfoot was that the boy had a huge

Bigfoot May Have Run Away From CC Camp Years Ago

Fri 10/17/58 Humboldt Standard

Bigfoot, the nocturnal wanderer of the Bluff Creek area, may be a giant youth who ran away from a CCC camp at Salyer in 1933.

Joseph Makely of Lutz, Florida, stated in an interview with the Tampa Tribune, Tampa, Florida, he remembers a boy six feet, 10 inches tall who, within two months of his arrival at Salyer's CCC camp had attained the height of seven feet and gained 30 pounds in weight.

Makely rememberd the boys first name was John and believes he was an orphan sent to the camp from San Francisco.

Because of his size and ungainly actions the boy was "razzed" by the other CCC men and finally left the camp, never returning.

"After all that ill treatment," Makely said, "John sort of lost his mind. He ran away into the forest and never came back. We saw him from time to time, as we worked on the fire trails, and it was rumored the Indians were caring for him, but John never came in for his pay."

"Now I am glad John has managed to survive, if he and Bigfoot are one and the same. I feel thankful to the Indians who cared for him and watched him grow into a legend," Makely added.

Whether or not Bigfoot is real or a hoax Makely felt it was a lesson in point for tolerance and said "I feel ashamed of myself for having been one of those who drove him out of his mind. It is sometimes too easy to do what the others do."

On October 17, 1958 the *Humboldt Standard*, the main paper for Humboldt County at the time, ran this article on their front page foot and appeared to be roaming the countryside. It should be noted that it's easily 30 miles of very rough country from Salyer to Bluff Creek.

Remember what Corky said to Tane when he picked her up after the sketch. He stated that her sketch was of an Indian. Nobody ever knows what happened to the boy that left the CCC camp. Could he account for the human DNA that might be in Bigfoot? Could the children that were purportedly left in the mountains, according to Ms. Peters, have played some role in Bigfoot's appearance having a human quality? Has Bigfoot always had a human look and quality, and we are now just discovering this? Are there multiple genetic lines to the Bigfoot species, and we have documented two of those through Harvey's sketches?

What's Next?

I've already started to interview witnesses in a new sighting area. When a good witness group is located I will again bring Harvey out for sketching. It will be interesting to see if the sketches in a completely different area, outside of the Hoopa region, will have the same human features as the sketches in Hoopa. I am trying to get the next area completed prior to any public release of Harvey's sketches, as I don't want to taint the witnesses. If the next area comes back with the same results as Hoopa, then I don't see the need to continue the same techniques as I've adopted here.

I have been very fortunate to be in a position to commit substantial time and resources to a cause in which I am deeply interested. Many of the volunteer researchers into Bigfoot can only commit some weekends and an occasional holiday. To truly make significant progress into the Bigfoot arena you need a substantial number of people over a 12-week summer. I would pick an area where they have a validated grouping of sightings and start high and work low. The Yurok call Bigfoot "Ridgewalkers" for a reason. Over the hundreds of years they have seen Bigfoot, they have continuously watched them walking the ridges. Why would they walk the ridges? I believe that it is an advantageous position to view possible prey, see movements of man, and is generally an area allowing quick movement, without a lot of bush, vines and large trees. It also offers a multitude of angles and avenues for escape.

It appears on certain occasions that Bigfoot makes its own temporary dens. I have personally documented several of these in this book, and I have also viewed these on film and other places in the wild. There are documented cases where people have stumbled onto Bigfoot sleeping in caves. It would seem likely that Bigfoot would take shelter under large boulders and burned out trees.

We know that certain animals in the wild go back to the location where they are born when they are ready to die. If Bigfoot is able to make it back to their place of birth, then we need to locate those places. Rather than spending nights on surveillance trying to see the creature moving, researchers should be spending time looking under boulders, in caves and in the hollow of logs, and maybe even in tree tops. All of this involves the luck of timing to find a deceased creature or the bones of a deceased Bigfoot. The more difficult the trip into the cave or under a boulder, maybe the more the likelihood of finding bones and other remains of a Bigfoot. If it is difficult for a researcher to get in, then it would be difficult for an animal to get in and take the remains. I also believe that the remains of a Bigfoot would more likely be found as soon as the trails are opened in a wilderness area. The people who are looking for remains need to be spending time in the appropriate areas. I've seen photos of a purported Bigfoot walking a ridgeline in the middle of winter. There has to be an instance when a Bigfoot was caught in an avalanche and subsequently died. The bottom of that ravine in the spring would be the place to find a corpse and or bones. There is probably a glacier somewhere in North America that harbors a Bigfoot corpse. With many of the glaciers at their smallest size in centuries, it may be a prudent place to start a search.

I sincerely hope that there isn't a hunter who would try to bring down a Bigfoot. There are many hunters who have claimed to shoot a Bigfoot; they hear it scream after the shooting, but NOBODY has ever brought back a body, ever. This creature is huge and it is doubtful that any caliber that is presently being used in the mountains of the United States would have the power to bring one down with one shot. This is not a challenge; please don't try to shoot a Bigfoot. I would be extremely leery of trying to shoot a creature of this size and not bringing it down immediately. The response by the creature may not bode well for the hunter.

If someone does shoot a Bigfoot and brings in a specimen, there may be a huge public response. As Skamania County has stated, this creature may be a humanoid and, as such, the hunter may be tried for murder. I have shown in this book that the creature has too many physical characteristics to ignore the obvious; it may be a relative to us. The creature has a great ability to avoid detection and capture. The creature can obviously think and rationalize under great pressure. It has the ability to adapt to changing conditions and environments as people move into their areas. There are several documented sightings that indicate that Bigfoot maintains some type of family structure. There are sightings where a large adult is seen with a smaller adult and then with a child. This sounds like a family to me.

During my research I have come across many sightings reports that emphasize the fact that Bigfoot has a human quality to it.

BFRO Website

Report #8547
2004
April 3
Spring
Panola County Texas

The summary of the sighting is that an individual was hog hunting from a stool he had placed high up in a tree. He was watching hogs in the area when the Bigfoot walked up. He watched the scenario unfold in front of him for 20 minutes. He had one of the longest looks at Bigfoot that anyone has claimed. Here are some of the statements made in the report.

> I tried to keep the scope on its head in case it came after me. It looked at me, cocked its head to one side. Its face had a curious expression, as if studying me. It opened its mouth showing its teeth, letting out a soft (RRRRRRR) sound. It then turned back to pick up the hog and walked off in no hurry as if I didn't exist.

> It was seven or eight feet tall, standing like a human on two legs, covered in hair and intensely watching the hogs. Its hands

> and feet looked human. It had breasts. Its ears seemed small for the size of its head.
>
> The eyes had no white they were solid brown.
>
> It had no waistline; waist was the same large bulk as the upper body. It had flat teeth like a human, no fangs. Face was human looking; it showed expression while looking at me.
>
> Its hair was shorter, about three inches long, just long enough to flap when moved.

Report #1372
1970
October 15
Fall
Routt County, Colorado

The summary of the sighting is that there were three men elk hunting in rural Colorado. They were walking back to camp after an unsuccessful day of hunting. As they were walking back, they heard sounds behind them. They thought they had spooked an elk, so they chased the creature making the sound. Portions of their statement follow.

> The back of the head appeared to be hunched slightly forward and without apparent neck visible. The back of the head was round skulled with no dome shape visible.
>
> I clearly observed the facial features through the scope. The face was flat, small mouthed and thin lipped. The nose was flat and large flared nostrils. The eyes appeared to be intelligent, what I would call a knowing look, black slightly wide apart and set below a low brow bridge, The face seemed to be lighter in color, possibly tan or light brown and hairless with a leathery appearance.

Report #6651
August 1982
Summer
Siskiyou County, CA
Location: South on Highway 96 from Somes Bar Store

[This next sighting was made just outside Orleans on the way to Dillon Creek, very near the road that comes off Highway 96 on the way to Forks of Salmon, and also very near the Damon Colegrove sighting.]

The summary of encounter is that the witness is a Karuk tribal member who was walking along the highway at approximately 6:00 p.m. The witness saw the Bigfoot stand up among a group of bushes on the opposite side of the road. Here are some of the statements contained in the report.

> I saw what appeared to be a big, hairy, tall man. I watch him for probably 30 seconds.
>
> It was covered in dark brown or black hair. I estimate it to be nine to 10 feet tall and its weight up to 900 or 1000 pounds.
>
> It appeared to have a bony face and the eyes were set deep and dark.
>
> The bony face stuck out because the creature's hair was all around its face.
>
> The report taker asked the witness what he thought the creature was, and his comment was, "They looked more human than anything else."

The sightings that are listed above show a consistency with the Hoopa sightings and witness statements. The color of the face in the second sighting is an unusual feature for a witness to see because most cannot get that close. The consistency of the statement regarding the teeth and the size of the ears also fall into line with Hoopa descriptions.

It may take many years of forensic sketches by qualified artists across

North America to document a significant cross section of Bigfoot, but I do believe that we have shown that there is a human quality to the creature.

Bigfoot has physical features that have a human quality but they also have communication styles, screams and other sounds that can be construed as human. There are also many documented cases (Pliny McCovey) where the witnesses believe they hear two Bigfoot communicating with some type of language. The language has been described in many different ways, but the consistent part of each story is that one Bigfoot talks, the other listens and then the other talks, very similar to a human interaction. If Bigfoot has its own language then it has a higher level of intelligence than many acknowledge. Several groups of researchers from throughout the United States have recorded purported Bigfoot screams. The screams appear to be a method to communicate and locate each other for a possible meeting. Phil Smith described this activity at the Bluff Creek Resort.

Many Native Americans from the Hoopa and Yurok Tribes believe that Bigfoot can understand their language, and because of that they have talked to it in their native language. Some of the witnesses believe that Bigfoot understood what they said. If Bigfoot does understand Hoopa or Yurok or even Kurok, could it be possible that it is a distant relative? Why would it understand a Native American language? Does the Native American boy who fled the CCC camp play some role in Bigfoot knowing the language? Every Native American tribe in Northern California, and many throughout the Untied States, has claimed a special relationship with Bigfoot. That bond has been described as spiritual, supernatural and sometimes just special, but it has always been reinforced in communications with elders. There is some type of long lasting mutual respect that exists between the Native Americans and Bigfoot and this is something that our research will continue to explore and attempt to understand. It is too strong a relationship to ignore.

Communicating, a family environment, using tools, advanced skills to avoid detection, and the ability to adjust behavior and living environment so it continues to evade capture are all traits of Bigfoot. I believe that all of these attributes lead to a conclusion that these creatures are intelligent. I can't name another creature in the wild that has been nearly as successful in evading capture, long-term surveillance or being scientifically recognized as a species. The arguments that Bigfoot is an illusion associated with UFOs, or are the fruit of a tourist's imagination gone wild, are over. Dreams and illusions do not leave tracks or hair that cannot be identified through DNA analysis. Harvey Pratt cannot draw sketches from witnesses (people who have never met) and have the drawings look like they could be drawn from the same subject. Impossible!

I believe that the time has arrived that legislators in the state governments need to stop avoiding the obvious. Bigfoot exists, it has existed for thousands of years and we need to protect it. I cannot find a documented case where Bigfoot has caused injury to anyone, but I can see a time in the future where either accidentally or purposely Bigfoot habitat is slowly eliminated to the point that it is cornered and people could be injured. Bigfoot does not live by society's rules and isn't schooled in our behavior. As in all of society, I am sure there are Bigfoot excluded from their groups for improper behavior in the same way we send people to prison. If the path of that Bigfoot crosses the path of an innocent hiker who has invaded its neighborhood, issues could evolve and people could be injured. If our legislators acknowledged the creature's existence, society could be educated about the creature and accidents could be avoided. I believe that legislators have inadvertently (or maybe not) made some attempts to protect the creature by designating certain forests as wilderness areas; no vehicular traffic and only people on horseback or on foot are allowed access. I've actually heard some people in the Bigfoot community claim that the wilderness area concept was developed mainly as a method to contain Bigfoot and allow its existence without continual human contact. It's an interesting idea considering it's entirely plausible that the Bigfoot visiting Hoopa probably live in the Siskiyou, Trinity Alps or Marble Mountain Wilderness Areas.

The Future

As I was finishing this book I was looking into other areas in the northwest for future study and comparative analysis with Hoopa. As luck would roll my way, I did find another region that is showing many of the promising aspects that Hoopa holds. In one of my first days in the region I was contacted by a woman who told me about her son who had several contacts with Bigfoot. She explained that her son was a caretaker for a cabin and property very high up in the mountains. The owner wanted a retreat that didn't have roads or visitors, nothing but the very rural feel of country living. The owner was well off and wanted the cabin and property maintained, and thus gave the job to her son. I am purposely withholding names, locations and specifics because I do not have permission from the involved parties to release this information and I still have not delved into the heart of my research into that region. The woman did put me in contact with her son.

I called the woman's son by phone (I'll call him Roger). Roger stated that it is a long hike into the region and is surrounded by public property. He said that nobody wants to take the hike because it is so steep, and they purposely never wanted to make it easy for fear of visitors. He said that the area is very lush, lots of deer and elk, and there are few people willing to make the trek to visit his site. Roger claimed to have had 12 direct encounters with several Bigfoot over a 20-year span of caretaking the cabin. He stated that he would reveal to me one story that scared a friend.

He had a friend who wanted to visit him but didn't quite know how difficult it would be to get to the cabin. He said he picked his friend up in town, they made the drive to the trailhead, and then started their hike. They were halfway there when they started to hear bipedal sounds on the perimeter. There were a lot more bipedal sounds than he was used to hearing on his walks. In past hikes he had heard a maximum of two Bigfoot on the perimeter, but this time they heard quite a few. He said that he didn't want to let his friend know that he was getting concerned, so they kept walking. At one point his friend got exhausted, so they stopped to rest.

Roger claimed that as they rested, three or four Bigfoot stood up around them on all four sides at a distance of maybe one hundred feet. He had never been more scared in his life. The creatures made no noise, there was no smell, nothing. It appeared, after a short time, that the Bigfoot were just interested in looking at them. There was a small gap between two of the creatures and they casually backed away and allowed Roger and friend to continue to the cabin. Roger said that the remainder of the trip was uneventful.

They spent the night without anything occurring. Early the next morning his friend woke him and stated that he had to get out of the cabin.

He told Roger that he had to walk him out right then because he couldn't stay. Roger said that he obliged and they walked back to the car without any incident. Roger had tried to get his friend to visit again, but no way. I did ask Roger to explain the facial structure of the creatures he had seen. He stated that they weren't as animal like as people might think, and they were somewhat a cross between animal and human but more on the human side. That sold me on his story.

Roger claimed that the sightings are usually in a 4–5 month time span every year. He believes that the Bigfoot migrate short distances and follow the herds of animals as they move within zones. He said that several areas around the property have huge game trails, very wide and very high. (It was interesting he said high game trails.)

I am presently meeting with Roger and spending considerable time in the area of the cabin. I will have a full and complete report on this area and many others when the next book is released. I'm sure that Harvey Pratt will again play a prominent role in sketching the Bigfoot that Roger and many others from the area observed.

It does not appear that the number of Bigfoot sightings and incidents are diminishing. In researching the sighting reports, there do appear to be years with high numbers of reports and years with almost zero reports. As baby boomers continue to retire and buy their dream houses in the mountains, society will continue to push into the Bigfoot environment. It would seem only natural that as new homes and developments are built in the forested environment, people will spot more Bigfoot in their travels to and from the market and during their leisure time around the house. I spent a great deal of time becoming educated about the dangers that exist in the wilds of certain parts of North America; Bigfoot was never on that list. If our elected officials continue to keep their heads in the sand and chanting the "we don't have a body" mantra, we may some day have a body to discuss — a human body. Let's protect the creature now, and protect society along the way.

Pass Bigfoot legislation today.

David Paulides
www.nabigfootsearch.com
nabigfootsearch@yahoo.com
2008

Online Resources

I believe that information is power. The sites below are a comprehensive list of Bigfoot, Sasquatch, Yeti and Skunk Ape websites from around the world. We encourage our readers to browse the web and become educated on the facts surrounding this interesting creature. I don't agree with the viewpoints expressed on all the sites, but it does make for provocative conversation.

Nabigfootsearch.com

Bigfoot

Bigfootsightings.org
Bigfootbiologist.org
Bigfootmuseum.org
Bigfootsounds.com
Westcoast-sasquatch.com
Bigfootcountry.net
Ourbigfoot.com
Greencountrybigfoot.info
Internationalbigfootsociety.com
Mid-americanbigfoot.com
Tnbigfootlady.com
Georgiabigfoot.com
Michiganbigfoot.org
Oregonbigfoot.com
Texasbigfoot.org
Trueseekers.org
Virginiabigfootresearch.org
Bigfootinfo.org
Bigfootproject.org
Marylandbigfootresearch.org
Bigfootforums.com
Bigfootencounters.com
Bfro.net
Recentbigfootsightings.com
Hdbrp.com
Upbigfoot.com
Searchingforbigfoot.com
Carolinabigfoot.com
Kentuckybigfoot.com
Ohiobigfoot.org
Easttexasbigfoot.com
Texasbigfoot.com
Pabigfootsociety.com

Skunk Ape

Skunkape.info

Sasquatch

Rfthomas.clara.net
Gcbro.net
Teamnesra.net
Sasquatchonline.com
Sasquatch-pg.net
Sasquatch-bigfoot.com

Canada

Manitobasasquatchsociety.com
Skookumquest.com
westcoast-sasquatch.com
WW1.freewebs.com/casr
Wriversasquatchassoc.net
Ontariosasquatch.com
Bcscc.ca

Australia

theaustralianyowieresearchcenter.com
Yowiehunters.com
Legendofyeti.com

Bigfoot Museums

Bigfootmuseum.com
Bigfootdiscoveryproject.com

Media

Coasttocoastam.com
Rense.com

Index